LESSONS
LEARNED

LESSONS LEARNED

What International Assessments
Tell Us about Math Achievement

TOM LOVELESS
editor

BROOKINGS INSTITUTION PRESS
Washington, D.C.

Library of Congress Cataloging-in-Publication data
Lessons learned : what international assessments tell us about math achievement /
Tom Loveless, editor.
    p.   cm.
  Summary: "Examines student performance in mathematics, using the Trends in
Mathematics and Science Study database to address pressing questions about school
policy and educational research, including major lessons learned from TIMSS test-
ing, comparison of math curricula across nations, effect of technology in the class-
room, and progress of popular math reforms abroad"—Provided by publisher.
  Includes bibliographical references and index.
  ISBN-13: 978-0-8157-5334-6 (cloth : alk. paper)
  ISBN-10: 0-8157-5334-9 (cloth : alk. paper)
  ISBN-13: 978-0-8157-5333-9 (pbk. : alk. paper)
  ISBN-10: 0-8157-5333-0 (pbk. : alk. paper)
  1. Mathematics—Study and teaching—Evaluation—Databases. 2. Mathematical
ability—Testing—Databases. 3. Mathematics teachers—Evaluation—Databases. 4.
Science—Study and teaching—Evaluation—Databases. 5. Science teachers—
Evaluation—Databases. 6. Science—Ability testing—Databases. 7. International
Association for the Evaluation of Educational Achievement—Examinations. I. Loveless,
Tom, 1954– II. TIMSS International Study Center.
  QA135.6.L47 2007
  510.71—dc22   2007026952

Typeset in Adobe Garamond

Composition by Circle Graphics
Columbia, Maryland

Printed by R. R. Donnelley
Harrisonburg, Virginia

Contents

Foreword

As a think tank with a "dot.edu" rather than "dot.org" domain name in cyberspace, Brookings takes especially seriously its commitment to work on educational issues. We do so largely through the work of our Brown Center and the expertise of Tom Loveless. I am proud to introduce the latest in a long series of high-quality, high-impact products of the Brown Center.

The first international study of student academic achievement was carried out between 1961 and 1965. Since then, several studies of math, reading, science, and civics have been conducted. Results are used to compare countries, but rarely is there further analysis of the data to explain how different nations actually educate their populations. Moreover, rarely do studies look beyond national rankings to address issues being debated all over the world. Questions concerning instruction—are some ways of teaching better than others?—and curriculum—what should students learn at different grade levels?—and policy—are small classes always more effective than large classes?—lend themselves to deeper, more sophisticated studies of international achievement data.

The International Association for the Evaluation of Educational Achievement (IEA) periodically administers the Trends in International Mathematics and Science Study (TIMSS). Fortunately, in addition to measuring achievement, IEA collects a wealth of TIMSS data from students, teachers, and school administrators. Yet these data are rarely mined for the insight they might provide.

In November 2006 Brookings Institution hosted a conference to further probe the TIMSS database. Scholars, educators, and policymakers from around the world—160 in all—convened to present research using TIMSS and other international testing data. The objective was to produce new knowledge on topics in curriculum, instruction, and policy by using secondary analyses of TIMSS data. Researchers presented findings on three main themes: curriculum (what students are taught, the skills and knowledge acquired in school); instruction (how teachers teach and the pedagogical strategies employed by teachers); and schools and school systems (including policy questions pertaining to the organization of schools and school systems).

This book is a collection of papers on math achievement that were presented at the conference. Topics are wide ranging. From a historical perspective, some authors examine what the international community has learned from four decades of international assessments and the differences between various international testing regimes. Others examine mathematics curricula: how the U.S. mathematics curriculum compares overseas, what have been curriculum reforms here, and how algebra is taught in the United States and abroad. Other authors examine student achievement: whether the relationship between achievement and school size holds up internationally, what is the link between technology and math achievement, and how student and classroom level factors over time relate to achievement. Each chapter is a unique and important contribution to international educational research.

Since the nineteenth century, advocates of common schooling in the United States have visited abroad and brought home ideas about what and how students should learn in classrooms. Education, schools, and achievement are viewed through an international lens even more so today. As the international community awaits the results of the 2007 TIMSS test, we can look forward to more data with which to conduct rigorous secondary analyses. We can go beyond the international student-achievement horse race and see what works in and out of the classroom to improve student learning in mathematics.

STROBE TALBOTT
President

September 2007
Washington, D.C.

Acknowledgments

Both the conference and book were supported by the National Science Foundation under grant 0608013, "Proposal for an International Conference on Mathematics: Analyses of IEA Studies in Mathematics; November 9–11, 2006; Washington, D.C." Additional support from the National Center for Education Statistics is gratefully acknowledged. At Brookings Institution Press, Marty Gottron and Starr Belsky copyedited the manuscript. The editor finally wishes to thank Katharyn Field-Mateer and Gladys Arrisueño for their assistance in planning and managing the conference.

1

Introduction: Secondary Analysis of Data from International Assessments

TOM LOVELESS

International tests in mathematics are famous for their horse race appeal: to find out the highest-scoring countries, the lowest-scoring countries, and for the citizens of any particular country, how its students rank against the rest of the world. Much more can be learned from analyzing the vast amount of data collected in international assessments. This book examines data from the math tests administered by the International Association for the Evaluation of Educational Achievement (IEA). In the chapters assembled here, experts on international assessment dig beneath the national scores to uncover evidence relevant to policy and practice.

In the first chapter, Ina Mullis and Mick Martin describe what has been learned from the four decades of IEA tests in math. They start with some history. The birth of international testing can be traced to the late 1950s, when Benjamin Bloom, a renowned scholar at the University of Chicago, drew up plans for a cross-national study of mathematics achievement. The First International Mathematics Study was conducted in twelve countries from 1961 to 1965. By producing credible results that attracted the interest of education researchers worldwide, the project quieted concerns about the feasibility of comparing student achievement of different countries. The Second International Mathematics Study took place from 1980 to 1982 and included several measures of curriculum and instruction. In addition to investigating curricular and instructional factors that may influence learning, conducting a second test eased lingering doubts and institutionalized the legitimacy of international testing.

In 1994–95, the Third International Mathematics and Science Study (TIMSS) brought science and mathematics together and introduced a number of psychometric innovations to international testing (such as constructed-response items, matrix sampling, scaling based on item-response theory, and multiple-imputation methods). In 1999 the TIMSS repeat was given to eighth graders, establishing a pattern of four-year intervals between tests. In 2003 the acronym TIMSS was kept, but the test's name was changed to the Trends in International Math and Science Study. Both fourth and eighth graders were examined, and plans were made for the next round of testing in 2007.

What are the main lessons that have been learned? Mullis and Martin present three findings that frame the chapters that follow. First, math achievement varies broadly among nations. At the eighth-grade level in 2003, for example, the average student in the top-scoring nation of Singapore scored higher than the ninety-fifth percentile in the low-scoring nations of Botswana, Ghana, Saudi Arabia, and South Africa. The second lesson is that high-scoring nations succeed with virtually all of their students. In the four or five top-scoring Asian nations, only between 1 and 4 percent of students do not reach minimal levels of competency. In the United States, about 10 percent of students do not reach minimal levels. The performance of the highest-scoring nations belies the notion that widespread excellence is impossible or that high test scores are largely driven by an inordinate focus on high achievers.

The third major lesson pertains to the national factors that promote achievement. Wealth matters. High-scoring nations tend to be wealthier, but the relationship is not ironclad. Korea, Chinese Taipei, and many of the former Eastern Bloc nations have modest gross national income statistics but still score relatively high on TIMSS. Cultural support for learning mathematics appears instrumental in promoting high achievement. In all countries, an emphasis on striving seems to be reinforced in the home of successful students. What is taught and studied matters. Not surprisingly, nations that report greater coverage of the topics tested on TIMSS tend to score higher than those with less coverage. Keeping students enrolled in school for longer periods of time—near-universal secondary education became a worldwide reality since the first IEA study—and providing a rigorous curriculum to all students underscores the necessity of what is now known as "opportunity to learn." As Mullis and Martin point out, TIMSS continues to collect an enormous amount of information on curriculum, instruction, teacher characteristics, and school organization, and these data have tremendous potential for shedding light on the many complex questions facing mathematics education.

Cross-national data run into methodological problems when it comes to testing causal hypotheses. Despite this, Jan-Eric Gustafsson is optimistic that international assessments can produce useful information for policymakers. In his chapter, Gustafsson acknowledges that differences among countries, especially in terms of the countless unmeasured influences on student achievement, make cross-sectional, national test scores a weak source of evidence to evaluate causal claims. Social scientists point to bias induced by endogeneity, omitted variables, and selection effects as methodological problems that are inherent in comparative studies and mostly impervious to the regression techniques that researchers use to mitigate them. Gustafsson argues that comparisons of within-country changes over time can address some of these methodological challenges.

Gustafsson illustrates the approach with two examples. He first looks at the relationship of chronological age and achievement using data from the 1995 and 2003 TIMSS tests. How much does age correlate with math learning? Within each of the cross-sectional data sets, no relationship is evident. But analysis of longitudinal changes in variables yields a different conclusion. Gustafsson first computes the change in age of each country's eighth graders who took the test and then investigates whether these changes correlate with changes in math scores. The correlation coefficient is .58 ($p < .05$). Nations that saw the greatest gains had older students taking the tests in 2003 than in 1995. The unstandardized regression coefficient (38) implies that a change of one year in age is associated with an increase in achievement of thirty-eight points. The direction of the effect (age and achievement are positively correlated) and magnitude are similar to previous studies.

Gustafsson applies the same technique to the relationship of class size and achievement. He finds that small fourth-grade classes are positively correlated with achievement, but no significant relationship can be detected at the eighth-grade level. This finding is consistent with other experimental studies, such as the STAR experiment in the United States, which show young children benefiting from small classes but the effect dissipating with older students. The methods Gustafsson employs to examine these two policy interventions could be used with other policy proposals using TIMSS data.

In the 1990s William Schmidt and other researchers used the TIMSS data to describe the U.S. math curriculum as "a mile wide and an inch deep." The argument was that some nations, the United States in particular, attempt to cover so many different topics that students learn only superficial concepts and are not given the chance to learn math concepts in depth. Schmidt and Richard T. Houang extend this line of thinking in "Lack of Focus in the Mathematics Curriculum: Symptom or Cause?" The charge pertains to coherence and focus. Coherence addresses when topics are introduced, taught, and then extinguished from the curriculum. Focus refers to the number of topics covered. A coherent and focused curriculum teaches only essential topics that are arranged sequentially to reflect the underlying structure of the discipline. The objective of the chapter is to find out if coherence and focus are related (or perhaps even redundant) and whether they are, either alone or together, related to achievement. Do nations with coherent and focused curricula outperform nations in which the curriculum covers more topics or is organized chaotically?

Schmidt and Houang run regression models using 1995 TIMSS data from grades three, four, seven, and eight to estimate the effect of coherence and focus on national achievement. Coherence is related to achievement in grades seven and eight, but falls short of statistical significance in grades three and four. Focus is not related to achievement at any grade level. That changes, however, in models that include both coherence and focus. Controlling for coherence, the amount of focus in the curriculum becomes statistically significant in its impact on achievement at all four grade levels. Schmidt and Houang conclude that the United States, in particular, should attempt to focus on fewer math topics arranged in a sequential manner, and they note the effort of the *Curriculum Focal Points for Prekindergarten through Grade 8 Mathematics*, published by the National Council of Teachers of Mathematics, to offer guidance on how American curricula can accomplish that task.

Jeremy Kilpatrick, Vilma Mesa, and Finbarr Sloane look at U.S. algebra performance. Beginning in the first and second grades, American mathematics textbooks contain lessons on content described as "algebra," although on close examination, the books typically offer little more than practice at discerning numerical patterns. In middle and high schools, the United States is unique in offering students a separate course called "algebra." In most countries, algebra is covered within courses that also feature geometry and other mathematical concepts. About one-third of American students have completed a full-year algebra course by the end of eighth grade. Many eighth graders in the United States are still studying arithmetic. How do American students do on the portion of TIMSS addressing algebra proficiency?

To answer this question, Kilpatrick and his colleagues analyze how U.S. students perform on algebra items found on TIMSS, focusing on the most difficult and easiest items on the test. They conclude that U.S. students perform at approximately the same level in algebra as they do across all domains of mathematics. The analysis identifies strengths and weaknesses. Fourth graders do relatively well on items requiring straightforward calculations and reasoning. Twelfth graders are better at interpreting the graph of a function than at other skills, but compared with their peers in other countries, their performance is weak.

American eighth graders diverge from the rest of the world in several areas. They are fairly good at understanding the notation for exponents, interpreting simple algebraic expressions, reasoning about sequences, and solving equations in which one variable is expressed in terms of another. They are relatively weak in interpreting symbols in an equation, completing a table showing a relation between variables, identifying a pattern, manipulating a simple algebraic expression, solving word problems involving relations between quantities, and translating words into algebraic symbols. In particular, Kilpatrick, Mesa, and Sloane express dismay at U.S. eighth graders' lack of proficiency in setting up equations to model real situations, despite the fact that eighth-grade math teachers in the United States report spending a lot of classroom time relating mathematics to daily life. The authors urge more opportunities for students to work with complex problems.

Mathematics education made headlines in the United States during the 1990s. A powerful reform movement, championed by the National Council of Teachers of Mathematics, ran into ferocious resistance. Many of the press headlines were devoted to political battles fought over how math should be taught. The instructional activities embraced by math reformers arose from a philosophical camp known as educational progressivism and included group work, calculator use as early as kindergarten, frequent teaching with manipulatives, student-centered inquiry over teacher-led direct instruction, writing about math, and open-ended assessment items.

In the chapter "What Can TIMSS Surveys Tell Us about 1990s Mathematics Reforms in the United States?" Laura S. Hamilton and José Felipe Martínez ask whether TIMSS data can shed light on the effectiveness of these reforms. They examine the eighth-grade TIMSS data from four countries—the United States, Japan, the Netherlands, and Singapore—in 1995, 1999, and 2003. Hamilton and Martínez find that American teachers, unlike teachers in the other three countries, significantly changed their practice from 1995 to 2003; however, it is possible that the differences detected are more because teachers changed how they discern and report instructional practices than because of any substantive changes in the practices themselves.

To investigate the impact of instruction on student achievement, Hamilton and Martínez construct an index to reflect several reform-oriented instructional strategies. What do they find? The cross-sectional analysis of 2003 achievement data does not reveal a strong relationship between more frequent use of reform-oriented practice and student achievement. Hamilton and Martínez urge readers to interpret these findings cautiously and discuss three challenges for future research on this topic: finding better ways to measure instructional practice, gathering more complete information on how instructional reforms are implemented, and conducting studies with an experimental design or a design that allows for causal inferences to be made.

Gabriela Schütz tackles the issue of school size. A worldwide movement in support of smaller schools has recently materialized, arguing that schools have grown too large, becoming too impersonal and too bureaucratized as institutions. Researchers theorize that an optimal school size may exist, that schools should be large enough to offer students an array of curricular choices while at the same time not becoming so big as to sacrifice their intimacy and other humane qualities. American researchers have pegged the optimal size as 600–900 students. Schütz mines the 2003 eighth-grade TIMSS data to see whether a relationship exists between school size and student achievement within different countries, and, she inquires, if such a relationship does exist, who profits most with respect to students of varying socioeconomic and immigrant status.

Schütz finds that school size is unrelated to achievement in most countries once statistical controls for family background of students and teacher and school effects are put in place. For eleven of the fifty-one countries and regions in the analysis, larger schools generally outperform smaller schools. There appeared to be no difference in effects among students of different socioeconomic status. Reducing the size of large schools may produce nonacademic benefits for students, but in the TIMSS data at least, very little support can be found for the notion that smaller schools produce higher achievement.

Elena C. Papanastasiou and Efi Paparistodemou look at technology use and student achievement. In research with pre-2003 international data, researchers uncovered an inverse relationship between technology use and student achievement in mathematics, with more frequent use of technology associated with lower performance. Employing latent variable analysis, Papanastasiou and Paparistodemou investigate whether the same inverse relationship exists in 2003 TIMSS data. Technology use is defined as student use of computers or calculators in their math studies. The study includes student-level data on eighth graders in four countries that vary in technology use and achievement: Cyprus, the Russian Federation, South Africa, and the United States.

The results are not encouraging for those who believe technology will revolutionize learning. After controlling for student background (primarily with a composite variable that measures socioeconomic status), Papanastasiou and Paparistodemou find effects that are small and negative. More technology use is correlated with slightly lower test scores. They also find, as numerous researchers before them have discovered, that student background swamps technology in its effect on achievement. Students from wealthier households and with parents who are better educated outscore students from poorer families headed by parents with less education, and technology holds little promise of making this situation more equitable. Papanastasiou and Paparistodemou point out that perhaps the promise of technology as a teaching tool has not been realized because teachers do not know yet how to use it effectively. The authors note that a small positive effect is found for calculator use in the

United States, perhaps as a result of math reformers' push to use calculators in support of higher-order thinking.

In 2000 the first assessment of the Program for International Student Assessment (PISA) was administered, a project of the Organization for Economic Cooperation and Development (OECD). Unlike TIMSS, PISA is given to an age-based sample (fifteen-year-olds) and attempts to measure mathematical literacy, which is defined as competencies needed for adulthood and not necessarily rooted in the school curriculum. Many nations were confused by the conflicting signals sent by the two tests. Eastern European countries, for example, scored lower on PISA than their TIMSS test results predicted, and the opposite was true for Western European countries, the United States, and Japan. In their chapter, "Are We the Man with Two Watches?" Dougal Hutchison and Ian Schagen ask whether PISA and TIMSS are comparable.

Hutchison and Schagen find that the two tests are alike on several technical characteristics, despite the obvious difference that PISA surveys fifteen-year-olds and TIMSS surveys eighth graders. The two assessments employ similar sampling designs; both use item-response theory to scale test results; and both generate plausible values for each student on each scale. They both gather data on pupils' home background and the characteristics of schools, with TIMSS going one step further and also gathering information from classroom teachers.

It is the content of items—the type of knowledge that the two tests attempt to measure—that differentiates PISA and TIMSS. Hutchison and Schagen describe TIMSS as a test that assesses "What do you know?" and PISA as assessing "What can you do?" TIMSS focuses on school-based knowledge, whereas PISA examines how well students can apply what they know to real-world situations. TIMSS items are more likely to be in multiple-choice format and to require the reproduction of facts or the use of standard algorithms, while PISA items are more likely to be constructed response and require making connections between existing skills and knowledge. PISA test items involve more reading, and as a 2006 study from the United Kingdom pointed out, the mathematics needed to answer items may be lowered so that students can apply knowledge in new contexts. These differences are important to keep in mind when the next round of PISA and TIMSS scores are released.

As demonstrated in the eight chapters that follow, international assessments have much more to contribute than simple rankings of nations on math achievement. International data can be used to investigate important policy questions, to evaluate the impact of particular practices, and to explore why some school systems are more effective than others. With the importance of mathematics continuing to increase, these topics—which were once of interest only to the elites of a few nations—are now the concern of nearly everyone in the world.

2

TIMSS in Perspective: Lessons Learned from IEA's Four Decades of International Mathematics Assessments

INA V. S. MULLIS AND MICHAEL O. MARTIN

International comparative studies of educational achievement had their origins in what is now known as the First International Mathematics Study (FIMS). FIMS was the first venture of the fledgling International Association for the Evaluation of Educational Achievement (IEA), the organization that pioneered international assessments of student achievement. Although FIMS was conducted principally between 1961 and 1965, its origins may be traced to a plan for a large-scale cross-national study of mathematics presented by Professor Benjamin S. Bloom in 1958 to colleagues in England and Germany.[1]

The 1950s were a time of great educational development and expansion in many countries, with excellent descriptive studies of countries' education systems conducted under the auspices of, for example, UNESCO and the Organization for Economic Cooperation and Development. However, the absence of any way of comparing the productivity or outputs of education systems was keenly felt by educational policymakers. Although the methodology for measuring student achievement through national surveys was established in many countries, it had not been implemented in a cross-national study. In FIMS, this methodology was applied for the first time in an international study of student achievement—mathematics in this case.

The First International Mathematics Assessments

The overall aim of the FIMS researchers was to use psychometric techniques to compare outcomes of different education systems. An underlying idea was that the world could be considered as an education laboratory, with each national education system an instance of a natural experiment in which national education policies and practices are treated as inputs and student attitudes and achievement are treated as outputs.

Education was considered as part of a larger sociopolitical-philosophical system in which rapid change was taking place, and the most fruitful comparisons were considered to be those that took account of how education responded to changes in the society. One aim, therefore, was to study how such changes influenced mathematics teaching and learning. The project created hypotheses and systematically addressed questions in three major areas: school organization, selection, and differentiation; the curriculum and methods of instruction; and the sociological, technological, and economic characteristics of families, schools, and societies.

Mathematics was chosen as the subject of the first international study of student achievement for several reasons. First was the importance of the subject itself. Countries were concerned with improving science and technology, and these are based in a fundamental way on the learning of mathematics. Second, was the question of feasibility. Evidence from studies of the mathematics curriculum and the teaching of mathematics suggested that countries had similar curricula and approaches to teaching, and so there was a good chance that a mathematics test could be constructed that would be an acceptable match to countries' curricula. Furthermore, translation into the languages of instruction of participating countries was expected to be less problematic for mathematics than for other subjects because of the widespread acceptance of mathematical symbols and notation.

FIMS was conducted in twelve countries, with the results published in two volumes in 1967.[2] These volumes summarize a wide range of findings related to the structure and organization of the educational systems; the mathematics curriculum and how it is taught; the mathematics teaching force and how it is educated and trained; and the attitudes, home background, and, most important, the mathematics achievement of students in the target populations. The target populations for the study were the grade in each country with most 13-year-old students and those students at the end of secondary education who had taken advanced courses in mathematics. FIMS was important not only for the immense contribution it made in its own right to the comparative study of mathematics achievement, but also because of its pioneering role in establishing without doubt the feasibility of comparative study of student achievement across different countries with different languages and cultures. FIMS showed for the first time that international studies could produce data on educational contexts and student achievement (at least in mathematics) that could form the basis of policy research aimed at improving education around the world.

The Second International Mathematics Study

By the middle of the 1970s, IEA researchers had begun to think that the time had come for a second international study of mathematics (SIMS), one that would build on the successes of the first study and incorporate all that had been learned about international studies in the intervening period. Planning for SIMS began in 1976, with data collection taking place in 1980–82. Like its predecessor, SIMS was a comparative study of student achievement in mathematics. In addition, however, SIMS placed great emphasis on the curriculum as a major organizing and explanatory factor, and also on the mathematics classroom—how teachers go about teaching mathematics to their students.

The three-level curriculum model introduced by SIMS has continued to underpin IEA studies in mathematics and science to this day. This model views the mathematics curriculum in a school system as having three aspects, each associated with a different level of the system. At the national or system level, the *intended curriculum* comprises the curricular goals and intentions that the country has for its students. These often are articulated through official documents such as national curriculum guides, course syllabi, and prescribed textbooks. The intended curriculum is what the ministry of education or similar responsible body expects to be taught in classrooms.

At the school or classroom level, the *implemented curriculum* comprises what is actually taught to students. Although, in an ideal world, the intended curriculum and the implemented curriculum may align closely, in reality that may not be the case. The teacher in the classroom is the central agent in implementing the mathematics curriculum, and the choices that the teacher makes regarding textbooks or other instructional materials, emphases on particular topics, and general teaching approach all have fundamental implications for the implemented curriculum. Teacher preparation can also have far-reaching implications, since teachers who have not received preparation in teaching the intended curriculum in mathematics are scarcely in a position to teach it effectively.

The third aspect of the curriculum model pertains to the student. The *attained curriculum* refers to the mathematics that the student has learned and the attitudes that the student has acquired as a result of being taught the curriculum in school. The attained curriculum may be considered the ultimate outcome of the educational process—how much of the mathematics intended at the national level and taught in the classroom has the student actually learned? The match between the intended, implemented, and attained curricula was an important focus of the second mathematics study.

SIMS was a very ambitious study. Its results were summarized in a series of volumes published over several years.[3] The analysis of the intended curriculum provided a detailed picture of each participating country's coverage of the mathematics topics in the curriculum framework. Data on the implemented curriculum were collected from teachers through an innovative "Opportunity to Learn" questionnaire, which asked teachers to review the items in the SIMS mathematics test and indicate whether they had taught the mathematics needed to answer each item.

Student achievement in the twenty countries or educational systems that participated in SIMS was the focus of the second volume of results.[4] Unlike in FIMS, where much effort was made to investigate causal relationships between the inputs of schooling and outputs in the form of student achievement, the focus of the SIMS achievement report was on describing a broad picture of mathematics education around the world. This included detailed information on student achievement in the major mathematics content areas addressed by SIMS—arithmetic, algebra, geometry and measurement, and descriptive statistics for the junior population (the grade with most 13-year-olds); and algebra, geometry, and elementary functions and calculus for the senior population (students at the end of secondary school having taken advanced preparation in mathematics). SIMS also reported extensively on student attitudes, preferences, and opinions, as well as on teacher preparation and attitudes toward mathematics.

TIMSS: The Third International Mathematics and Science Study

IEA's Third International Mathematics and Science Study (TIMSS) was the first to bring together mathematics and science in a single study. Building on the traditions of FIMS and SIMS, as well as on IEA's two earlier studies of student achievement in science, TIMSS was the most ambitious IEA study to date. At the heart of TIMSS was a far-ranging assessment of student achievement in mathematics and science at three levels of the education system—grades three and four (the end of primary schooling), grades seven and eight (middle school or lower secondary), and twelfth grade (or final grade). The twelfth-grade assessment included tests of advanced mathematics and physics for students having taken advanced courses in these subjects, as well as tests of mathematics and science literacy for all students at this grade level.

TIMSS also included an in-depth analysis of mathematics and science curricula and an extensive investigation into home, school, and classroom contexts for learning.[5] To provide fresh insights into the ways the curriculum is taught in mathematics classrooms, TIMSS incorporated video studies of instructional practices in Germany, Japan, and the United States.[6] The Case Study Project, which also focused on Germany, Japan, and the United States, employed a quasi-ethnographic methodology that relied on the interaction of experienced researchers with families and teachers and on information obtained from school authorities and policy experts. Extensive information was obtained about four predetermined topics: national standards, teacher training and teachers' working conditions, attitudes toward dealing with differences in ability, and the place of school in adolescents' lives.[7]

The launching of TIMSS at the beginning of the 1990s coincided with an upsurge of interest in international studies. There was a growing acceptance that effective mathematics and science education would be a crucial ingredient of economic development in the increasingly knowledge-based and technological world of the future. An additional impetus came from the recent breakup of the Soviet Union, which resulted in a host of newly independent countries from Central and Eastern Europe eager to participate in a study that promised to provide data to guide the revitalization of their education systems. Almost fifty countries participated in TIMSS in one way or another, making it the largest study of its kind at that time. The first TIMSS data collection took place during the 1994–95 school year.

TIMSS differed from earlier IEA studies in another important respect: it was well funded, principally by the U.S. National Center for Education Statistics and the National Science Foundation. With this support, TIMSS introduced a number of innovations that enhanced its validity and reliability as an international assessment of student achievement.

One of the essential underpinnings of TIMSS was an ambitious and extensive set of curriculum frameworks for mathematics and science.[8] Designed to support both a comprehensive analysis of curriculum documents and textbooks and the development of a challenging and broad-based assessment of student achievement, the frameworks sought to improve the validity of the TIMSS assessments by specifying a wide range of mathematics and science content to be covered, employing a variety of cognitive skills, and recommending different kinds of testing formats. The TIMSS achievement tests included not only multiple-choice items, which had commonly been used in international studies, but also constructed-response items that required students to provide a written response. TIMSS constructed-response items included both short-answer questions and questions requiring an extended response. A performance assessment was a special component of the assessments at the fourth and eighth grade levels.[9]

The TIMSS international reports of mathematics achievement describe student achievement in mathematics in the participating countries and present a wealth of data about the social and educational context for learning in each country.[10] TIMSS results were widely disseminated across participating countries, and the impact on educational systems has been considerable.[11]

The drive toward valid and reliable measurement of student achievement in mathematics and science inevitably resulted in an assessment with far too many items to be administered in the time available to test any one student (ninety minutes). This difficulty was readily solved by adopting a matrix-sampling booklet design, whereby each student took just a part of the assessment in the ninety minutes available. In the matrix-sampling approach, the pool of assessment items is distributed across a number of student booklets, with each student responding to just one booklet. Although this booklet design solved the problem of administering the assessment, it created another problem: how to report student achievement in a range of mathematics and science content areas when just a few items in each area have been administered to each student.

In a major step forward for international assessments, TIMSS addressed this issue by implementing a complex psychometric scaling technique that was developed by the U.S. National Assessment for Educational Progress (NAEP) for use in its national surveys of student achievement. This approach combines item-response-theory scaling with conditioning and multiple-imputation methods to provide each student with estimated scores on the entire assessment even though the student has responded to just a part of it. With TIMSS, for example, this technique was used to assign each student scores in both mathematics and science overall, as well as in each of the TIMSS mathematics and science content areas, for example, number, algebra, geometry, data, physics, chemistry, biology, and earth science, regardless of the test booklet taken by the student. The student scores may be used to describe student achievement and also to analyze the relationship between achievement and the many contextual variables that are measured as part of TIMSS.

The techniques that TIMSS pioneered in the field of international student assessment may be used in any study facing the challenge of administering a student assessment that is much too large to be taken by a single student in the testing time available. The ability to combine a data-collection design that allows the assessment to be administered piecemeal with an analytic methodology that enables a student's scores on the entire assessment to be estimated from just the pieces of the assessment that the student took provides enormous flexibility to assessment designers. It means that subject matter specialists have great freedom to design wide-ranging assessments that do full justice to their assessment goals, secure in the knowledge that the resulting data can be reported in a meaningful way.

The TIMSS funding afforded unprecedented opportunities for development and analysis and reporting, but it also brought with it the responsibility of ensuring that all aspects of TIMSS were conducted to the highest standards, and that this high quality was documented at every step of the way.[12] IEA's studies had always applied state-of-the-art techniques to the development and implementation of its studies, but TIMSS was the first study with the resources to do this in a way that ensured high-quality comparable data from the participating countries. For educational data to be useful in an international arena, it is not enough for them to be of high quality from a national perspective, the data also must be highly comparable from country to country.

More than any previous international study of student achievement, TIMSS placed the issue of data quality and data comparability at center stage. TIMSS identified a range of crucial areas for data comparability and went to great lengths to establish high standards, to report steps taken to meet these standards, and to attach consequences to failure to meet the standards. Among the questions that TIMSS focused on in striving for comparative validity were the following:

—Is curriculum coverage comparable?

—Are target populations comparable?

—Was sampling of populations conducted correctly?

—Are instrument translations comparable?

—Were the achievement tests administered appropriately?

—Was scoring of constructed-response items conducted correctly?

—Are the resulting data comparable?

This insistence on "comparative validity" has become a hallmark of TIMSS and is one of the reasons that international comparative data on student achievement have become accepted as reliable instruments of educational policy analysis.

The TIMSS Assessments

The TIMSS results revealed a wide range of mathematics and science achievement both within and across countries. For several countries, TIMSS also revealed differences in relative performance across the grades. For example, in the United States, students performed just above the international average at the fourth grade but below the international average at the eighth grade. At the twelfth grade, average mathematics performance of U.S. students was among the lowest of the participating countries. These findings suggest that U.S. students do reasonably well in the primary grades, but that the longer they stay in school, the more they fall behind their peers in other countries.

Results such as these prompted the decision to administer TIMSS at the eighth grade again in 1999, four years after the original TIMSS administration in 1995. Since the cohort of students that was in fourth grade in 1995 would be in the eighth grade in 1999, TIMSS 1999 (or TIMSS-Repeat as it was known at the time) would allow countries to see if the relative performance of their eighth-grade students in 1999 reflected their performance when they were in the fourth grade in 1995, or whether the eighth-grade students of 1999 more resembled their eighth-grade compatriots in 1995.

In addition to the analysis of the fourth-to-eighth-grade effect, TIMSS 1999 for the first time in an international study provided a solid measurement of trends in student achievement from two points in time—eighth-grade students in 1995 compared with eighth-grade students in 1999. This was possible because about one-third of the items from TIMSS 1995 had been kept secure for use in future studies, and these were combined with items newly developed for 1999 to form the TIMSS 1999 assessment. Applying the TIMSS scaling methodology to the data from 1995 and 1999 and capitalizing on the items common to both data sets, TIMSS was able to measure eighth-grade mathematics achievement in both 1995 and 1999 on the same scale. This provided countries that participated in both assessments with a direct measure of how much average eighth-grade mathematics performance had improved or declined over the four-year period.[13]

With the success of TIMSS 1995 and TIMSS 1999, IEA committed to TIMSS as a major component of its core cycle of studies. Now renamed the *Trends in International Mathematics and Science Study,* TIMSS is dedicated to providing sound measurement of trends in student achievement in mathematics and science at fourth and eighth grades every four years. TIMSS 2003 successfully reported achievement at both grades, extending the TIMSS trend line from 1995 through 1999 to 2003 for eighth-grade students and from 1995 to 2003 for students in fourth grade.[14] TIMSS 2007, currently in the data-collection phase, will further extend both fourth- and eighth-grade trend lines to include 2007.[15]

First administered in sixteen countries as part of the TIMSS 1995 twelfth-grade assessment, TIMSS Advanced assesses school-leaving students with special preparation in advanced mathematics and physics. With a second data collection in 2008, TIMSS Advanced will permit countries that participated in 1995 to determine whether the achievement of students having taken advanced coursework has changed over time.[16]

International Comparisons in Mathematics Achievement

Since TIMSS 2003 represents the culmination of experience in assessing mathematics internationally, this section of the paper highlights results from TIMSS 2003 to provide a perspective about what has been learned about students' mathematics achievement.[17]

Variations between and within Countries in Mathematics Achievement

Figure 2-1 presents the TIMSS 2003 results for the forty-five countries at the eighth grade and twenty-five countries at the fourth grade that participated fully in TIMSS 2003, meeting all guidelines for high-quality sampling and data collection. Countries are shown in decreasing order of average scale score, together with an indication of whether the country average is significantly higher (⬦) or lower (▼) than the average of the TIMSS mathematics achievement scale, which was set at 500 when the scale was developed in 1995. This represents a reanalysis of the presentation in the *TIMSS 2003 International Mathematics Report*, which compared achievement to the average of the participating countries (467 at the eighth grade and 495 at the fourth grade). Using an approach dependent on participating countries has caused the international average to shift with each assessment, whereas the TIMSS mathematics achievement scale has a mean that is stable over time.

At the eighth-grade level, thirteen countries performed above average on the TIMSS mathematics achievement scale, seven countries, including the United States, performed about the same as the TIMSS scale average (average scale score 504), and twenty-five countries performed below average. The disproportionate number of countries performing below the scale average reflects the number of TIMSS 2003 countries facing economic and educational challenges. For example, countries with high values on the Human Development Index (HDI) provided by the United Nations enjoy long life expectancy, high levels of school enrollments and adult literacy, and a good standard of living as measured by per capita gross domestic product. Only twelve of the TIMSS 2003 countries have an HDI greater than 0.9 (on a scale 0 to 1.0): Australia, Belgium (Flemish), England, Israel, Italy, Japan, New Zealand, Norway, the Netherlands, Scotland, Sweden, and the United States. At the fourth-grade level, thirteen countries, including the United States, performed above average (average scale score 518), three countries performed about the same as the TIMSS scale average, and nine countries performed below average.

The TIMSS 2003 results make it clear that the range in mathematics achievement across and within countries is substantial. Even though performance generally differed very little between one country and the next higher- or lower-performing country, with so many countries from diverse cultural and economic backgrounds, the achievement range for the eighth grade was particularly large—from an average of 605 for Singapore to 264 for South Africa. At the fourth-grade level, achievement scores ranged from 594 in Singapore to 339 in Tunisia.

Figure 2-1 also shows a graphical representation of the distribution of student achievement within each country. The dark boxes at the midpoints of the distributions show the 95 percent confidence intervals around the average achievement in each country. Achievement for each country is shown for the twenty-fifth and seventy-fifth percentiles as well as for the fifth and ninety-fifth percentiles. Each percentile point indicates the percentage of students performing below and above

Figure 2-1a. *Distribution of Mathematics Achievement*

a. *Eighth Grade*

Countries	Average scale score	Mathematics achievement distribution
Singapore	605 (3.6) ⭗	
⊷ Korea, Rep. of	589 (2.2) ⭗	
†Hong Kong, SAR	586 (3.3) ⭗	
Chinese Taipei	585 (4.6) ⭗	
Japan	570 (2.1) ⭗	
Belgium (Flemish)	537 (2.8) ⭗	
†Netherlands	536 (3.8) ⭗	
Estonia	531 (3.0) ⭗	
Hungary	529 (3.2) ⭗	
Malaysia	508 (4.1) ⭗	
Latvia	508 (3.2) ⭗	
Russian Federation	508 (3.7) ⭗	
Slovak Republic	508 (3.3) ⭗	
Australia	505 (4.6)	
‡United States	504 (3.3)	
¹Lithuania	502 (2.5)	
TIMSS Scale Avg.	**500**	
Sweden	499 (2.6)	
†Scotland	498 (3.7)	
²Israel	496 (3.4)	
New Zealand	494 (5.3)	
Slovenia	493 (2.2) ▼	
Italy	484 (3.2) ▼	
Armenia	478 (3.0) ▼	
¹Serbia	477 (2.6) ▼	
Bulgaria	476 (4.3) ▼	
Romania	475 (4.8) ▼	
Norway	461 (2.5) ▼	
Moldova, Rep. of	460 (4.0) ▼	
Cyprus	459 (1.7) ▼	
²Macedonia, Rep. of	435 (3.5) ▼	
Lebanon	433 (3.1) ▼	
Jordan	424 (4.1) ▼	
Iran, Islamic Rep. of	411 (2.4) ▼	
¹Indonesia	411 (4.8) ▼	
Tunisia	410 (2.2) ▼	
Egypt	406 (3.5) ▼	
Bahrain	401 (1.7) ▼	
Palestinian Nat'l. Auth.	390 (3.1) ▼	
Chile	387 (3.3) ▼	
¹‡Morocco	387 (2.5) ▼	
Philippines	378 (5.2) ▼	
Botswana	366 (2.6) ▼	
Saudi Arabia	332 (4.6) ▼	
Ghana	276 (4.7) ▼	
South Africa	264 (5.5) ▼	

0   100   200   300   400   500   600   700

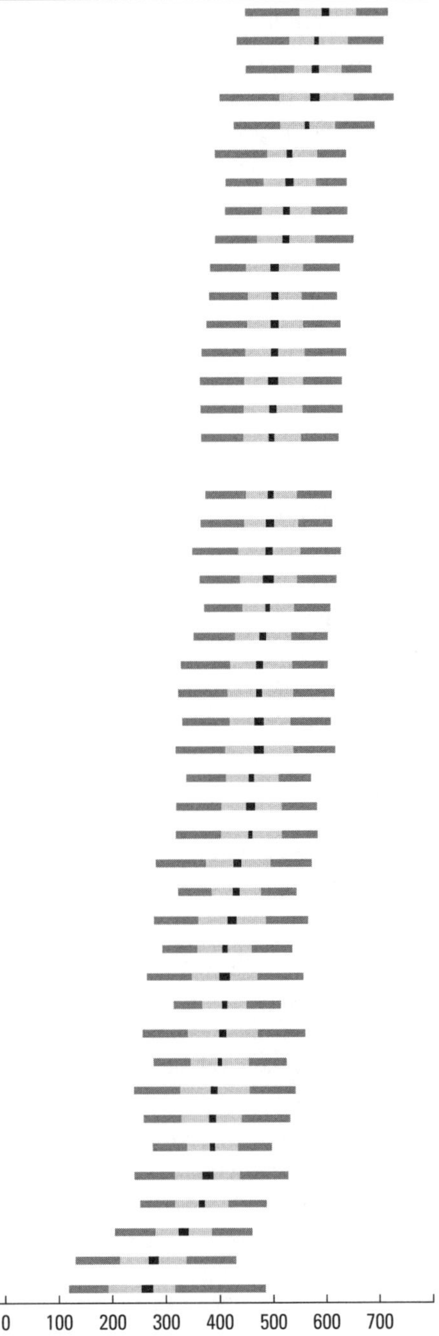

Figure 2-1b. *Distribution of Mathematics Achievement*

b. *Fourth Grade*

Countries	Average scale score	Mathematics achievement distribution
Singapore	594 (5.6) ⬤	
†Hong Kong, SAR	575 (3.2) ⬤	
Japan	565 (1.6) ⬤	
Chinese Taipei	564 (1.8) ⬤	
Belgium (Flemish)	551 (1.8) ⬤	
†Netherlands	540 (2.1) ⬤	
Latvia	536 (2.8) ⬤	
‡Lithuania	534 (2.8) ⬤	
Russian Federation	532 (4.7) ⬤	
†England	531 (3.7) ⬤	
Hungary	529 (3.1) ⬤	
†United States	518 (2.4) ⬤	
Cyprus	510 (2.4) ⬤	
Moldova, Rep. of	504 (4.9)	
Italy	503 (3.7)	
TIMSS Scale Avg.	**500**	
†Australia	499 (3.9)	
New Zealand	493 (2.2)	
†Scotland	490 (3.3) ⬇	
Slovenia	479 (2.6) ⬇	
Armenia	456 (3.5) ⬇	
Norway	451 (2.3) ⬇	
Iran, Islamic Rep. of	389 (4.2) ⬇	
Philippines	358 (7.9) ⬇	
Morocco	347 (5.1) ⬇	
Tunisia	339 (4.7) ⬇	

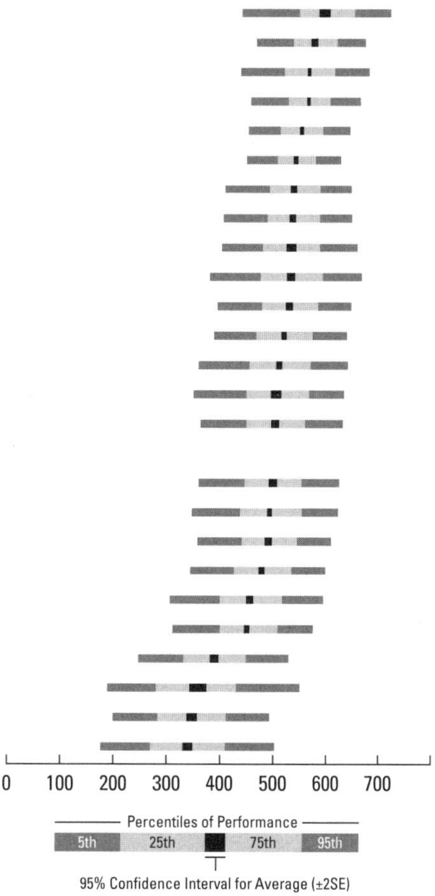

0   100   200   300   400   500   600   700

——— Percentiles of Performance ———

| 5th | 25th | ■ | 75th | 95th |

95% Confidence Interval for Average (±2SE)

Source: IEA's Trends in International Mathematics and Science Study (TIMSS).

† Met guidelines for sample participation rates only after replacement schools were included.

‡ Nearly satisfied guidelines for sample participation rates only after replacement schools were included.

1. National Desired Population does not cover all of International Desired Population.

2. National Defined Population covers less than 90% of National Desired Population.

►◄Korea tested the same cohort of students as other countries, but later in 2003, at the beginning of the next school year.

( ) Standard errors appear in parentheses. Because results are rounded to the nearest whole number, some totals may appear inconsistent.

⬤ Country average significantly higher than international average.

⬇ Country average significantly lower than international average.

that point on the scale. (For example, 25 percent of the eighth-grade students in each country performed below the twenty-fifth percentile for that country, and 75 percent performed above the twenty-fifth percentile.)

In most countries, the range of performance for the middle group of students (50 percent of students between the twenty-fifth and seventy-fifth percentiles) was about 100 to 130 scale points—not too dissimilar from the difference between performance in Singapore, on average, and the average of the TIMSS mathematics scale. The range of performance between the fifth and ninety-fifth percentiles (90 percent of the students) was approximately 270 to 300 points in most countries.

As well as showing the wide spread of student achievement within each country, the percentiles also provide a perspective on the size of the differences among countries. For example, average achievement for eighth-grade Singaporean students exceeded performance at the ninety-fifth percentile in the lower-performing countries such as Botswana, Ghana, Saudi Arabia, and South Africa. Similarly, average fourth-grade performance in Singapore exceeded performance at the ninety-fifth percentile in Iran, Morocco, the Philippines, and Tunisia. That difference means that only the most proficient students in the lower-performing countries approached the level of achievement of Singaporean students of average proficiency.

The TIMSS International Benchmarks of Achievement

Many countries are working toward the goal of all students achieving at high levels in mathematics. It is therefore interesting to look at mathematics achievement internationally from the perspective of how well countries are doing in having more students learn more mathematics. Considering achievement at four benchmarks along the TIMSS mathematics scale provides a way to consider this question, and the success of the top-performing Asian countries in reaching the benchmarks demonstrates that mathematics proficiency for all students within a country is possible.

Because the TIMSS mathematics achievement scale summarizes the performance of each country's students across a large number of test questions (about 170 at eighth grade and 160 at fourth grade), it is possible to examine achievement at various points along the scale in terms of the questions answered correctly by students performing at those scale levels (known as scale anchoring). Since the TIMSS test questions were designed to measure a wide range of student knowledge and proficiency in mathematics, analyzing student achievement at the four benchmarks provides a rich description of the mathematics that students know and can do at each benchmark. In analyzing performance across the test questions in relation to scores on the achievement scale at the eighth grade, it was determined that performance ranged from using relatively complex algebraic and geometric concepts and relationships at the advanced benchmark to having only some basic mathematical knowledge primarily in the area of number at the low benchmark.

While the full descriptions based on an ambitious scale-anchoring exercise conducted by international experts can be found in the *TIMSS 2003 International Mathematics Report,* achievement at the eighth-grade benchmarks can be summarized as follows:

—*Advanced* (a score of 625): Students can organize information, make generalizations, solve nonroutine problems, and draw and justify conclusions from data.

—*High* (550): Students can apply their understanding and knowledge in a wide variety of relatively complex situations.

—*Intermediate* (475): Students can apply basic mathematical knowledge in straightforward situations.

—*Low* (400): Students have some basic mathematical knowledge.

At the fourth-grade level, students achieving at or above the advanced benchmark showed the ability to solve a variety of problems (appropriate to fourth-grade curriculum). Fourth-grade students achieving at the low benchmark demonstrated an understanding of whole numbers and the properties of basic geometric shapes, and knew how to read simple bar graphs. Performance at the fourth-grade benchmarks is summarized below.

—*Advanced* (625): Students can apply their understanding and knowledge in a wide variety of relatively complex situations.

—*High* (550): Students can apply their knowledge and understanding to solve problems.

—*Intermediate* (475): Students can apply basic mathematical knowledge in straightforward situations.

—*Low* (400): Students have some basic mathematical knowledge.

Table 2-1 shows the percentage of students in each participating country that reached each international benchmark. Both the eighth- and fourth-grade results are presented in decreasing order by percentage reaching the advanced benchmark. The most striking finding is that such large percentages of students in the five top-performing Asian countries achieved at or above the advanced benchmark, especially at the eighth grade. Most remarkably, 44 percent of the Singaporean eighth-grade students scored at or above the advanced benchmark. Not only was their average performance the highest at 605, but a substantial percentage scored at or above 625, demonstrating proficiency, for example, in solving nonroutine problems in algebra and geometry. Four other East Asian countries also had large percentages of eighth-grade students achieving at or above the advanced level. The Republic of Korea and Chinese Taipei both had more than one-third of their students performing at the advanced

Table 2-1a. *Percentages of Eighth Grade Students Reaching TIMSS 2003 International Benchmarks of Mathematics Achievement*

Countries	Advanced international benchmark (625)	High international benchmark (550)	Intermediate international benchmark (475)	Low international benchmark (400)
Singapore	44 (2.0)	77 (2.0)	93 (1.0)	99 (0.2)
Chinese Taipei	38 (2.0)	66 (1.8)	85 (1.2)	96 (0.6)
▸Korea, Rep. of	35 (1.3)	70 (1.0)	90 (0.5)	98 (0.3)
†Hong Kong, SAR	31 (1.6)	73 (1.8)	93 (1.3)	98 (0.6)
Japan	24 (1.0)	62 (1.2)	88 (0.6)	98 (0.2)
Hungary	11 (1.0)	41 (1.9)	75 (1.6)	95 (0.8)
†Netherlands	10 (1.5)	44 (2.5)	80 (2.0)	97 (0.8)
Belgium (Flemish)	9 (0.9)	47 (1.9)	82 (1.2)	95 (0.9)
Estonia	9 (0.8)	39 (1.9)	79 (1.4)	97 (0.5)
Slovak Republic	8 (0.8)	31 (1.7)	66 (1.7)	90 (1.1)
Australia	7 (1.1)	29 (2.4)	65 (2.3)	90 (1.4)
‡United States	7 (0.7)	29 (1.6)	64 (1.6)	90 (1.0)
International Avg.	**7 (0.1)**	**23 (0.2)**	**49 (0.2)**	**74 (0.2)**
Malaysia	6 (1.0)	30 (2.4)	66 (2.1)	93 (0.9)
Russian Federation	6 (0.8)	30 (1.8)	66 (1.8)	92 (0.9)
2 Israel	6 (0.6)	27 (1.5)	60 (1.8)	86 (1.2)
Latvia	5 (0.7)	29 (1.5)	68 (1.7)	93 (0.8)
1 Lithuania	5 (0.6)	28 (1.2)	63 (1.4)	90 (0.8)
New Zealand	5 (1.3)	24 (2.7)	59 (2.5)	88 (1.7)
†Scotland	4 (0.6)	25 (2.1)	63 (2.4)	90 (1.1)
Romania	4 (0.6)	21 (1.8)	52 (2.2)	79 (1.7)
1 Serbia	4 (0.4)	21 (1.1)	52 (1.4)	80 (0.9)
Sweden	3 (0.5)	24 (1.2)	64 (1.5)	91 (1.0)
Slovenia	3 (0.5)	21 (1.0)	60 (1.3)	90 (0.9)
Italy	3 (0.6)	19 (1.5)	56 (1.7)	86 (1.2)
Bulgaria	3 (0.7)	19 (1.8)	51 (2.1)	82 (1.6)
Armenia	2 (0.3)	21 (1.3)	54 (1.5)	82 (1.0)
Cyprus	1 (0.2)	13 (0.7)	45 (1.0)	77 (1.0)
Moldova, Rep. of	1 (0.3)	13 (1.2)	45 (2.1)	77 (1.7)
2 Macedonia, Rep. of	1 (0.2)	9 (1.0)	34 (1.7)	66 (1.7)
Jordan	1 (0.2)	8 (1.0)	30 (1.9)	60 (1.9)
1 Indonesia	1 (0.2)	6 (0.7)	24 (1.7)	55 (2.4)
Egypt	1 (0.2)	6 (0.5)	24 (1.2)	52 (1.7)
Norway	0 (0.2)	10 (0.6)	44 (1.6)	81 (1.2)
Lebanon	0 (0.1)	4 (0.6)	27 (1.8)	68 (1.9)
Palestinian Nat'l. Auth.	0 (0.1)	4 (0.4)	19 (1.2)	46 (1.5)
Iran, Islamic Rep. of	0 (0.2)	3 (0.4)	20 (1.1)	55 (1.4)
Chile	0 (0.1)	3 (0.4)	15 (1.2)	41 (1.8)
Philippines	0 (0.1)	3 (0.6)	14 (1.7)	39 (2.7)
Bahrain	0 (0.0)	2 (0.2)	17 (0.7)	51 (1.1)
South Africa	0 (0.1)	2 (0.6)	6 (1.3)	10 (1.8)
Tunisia	0 (0.0)	1 (0.3)	15 (1.1)	55 (1.6)
‡Morocco	0 (0.0)	1 (0.2)	10 (0.9)	42 (1.6)
Botswana	0 (0.0)	1 (0.2)	7 (0.7)	32 (1.5)
Saudi Arabia	0 (0.1)	0 (0.1)	3 (0.6)	19 (1.7)
Ghana	0 (0.0)	0 (0.0)	2 (0.5)	9 (1.3)

Table 2-1b. *Percentages of Fourth Grade Students Reaching TIMSS 2003 International Benchmarks of Mathematics Achievement*

Countries	Advanced international benchmark (625)	High international benchmark (550)	Intermediate international benchmark (475)	Low international benchmark (400)
Singapore	38 (2.9)	73 (2.4)	91 (1.3)	97 (0.6)
†Hong Kong, SAR	22 (1.7)	67 (2.0)	94 (0.7)	99 (0.2)
Japan	21 (0.8)	60 (1.0)	89 (0.7)	98 (0.3)
Chinese Taipei	16 (0.9)	61 (1.1)	92 (0.7)	99 (0.2)
†England	14 (1.4)	43 (1.8)	75 (1.6)	93 (0.8)
Russian Federation	11 (1.6)	41 (2.6)	76 (2.0)	95 (0.8)
Belgium (Flemish)	10 (0.6)	51 (1.3)	90 (0.8)	99 (0.3)
Latvia	10 (0.9)	44 (1.9)	81 (1.3)	96 (0.7)
†Lithuania	10 (1.1)	44 (1.7)	79 (1.3)	96 (0.7)
Hungary	10 (1.0)	41 (1.6)	76 (1.6)	94 (0.8)
International Avg.	**9 (0.2)**	**33 (0.3)**	**63 (0.3)**	**82 (0.2)**
Cyprus	8 (0.7)	34 (1.2)	68 (1.2)	89 (0.7)
†United States	7 (0.7)	35 (1.3)	72 (1.2)	93 (0.5)
Moldova, Rep. of	6 (1.0)	32 (2.1)	66 (2.1)	88 (1.5)
Italy	6 (1.0)	29 (1.8)	65 (1.7)	89 (1.1)
†Netherlands	5 (0.8)	44 (1.5)	89 (1.2)	99 (0.4)
†Australia	5 (0.7)	26 (1.7)	64 (1.9)	88 (1.3)
New Zealand	5 (0.5)	26 (1.2)	61 (1.3)	86 (1.0)
†Scotland	3 (0.4)	22 (1.4)	60 (1.6)	88 (1.2)
Slovenia	2 (0.4)	18 (1.0)	55 (1.5)	84 (1.0)
Armenia	2 (0.3)	13 (1.2)	43 (1.7)	75 (1.5)
Norway	1 (0.2)	10 (1.0)	41 (1.3)	75 (1.2)
Philippines	1 (0.7)	5 (2.1)	15 (2.7)	34 (2.6)
Iran, Islamic Rep. of	0 (0.1)	2 (0.3)	17 (1.3)	45 (2.2)
Tunisia	0 (0.1)	1 (0.3)	9 (1.0)	28 (1.7)
Morocco	0 (0.0)	1 (0.2)	8 (0.8)	29 (2.2)

Source: IEA's Trends in International Mathematics and Science Study (TIMSS) 2003.

† Met guidelines for sample participation rates only after replacement schools were included.

‡ Nearly satisfied guidelines for sample participation rates only after replacement schools were included.

1. National Desired Population does not cover all of International Desired Population.

2. National Defined Population covers less than 90% of National Desired Population.

◄ Korea tested the same cohort of students as other countries, but later in 2003, at the beginning of the next school year.

( ) Standard errors appear in parentheses. Because results are rounded to the nearest whole number, some totals may appear inconsistent.

benchmark, 38 and 35 percent, respectively, followed by Hong Kong SAR (Special Administrative Region of China) with 31 percent. About one-fourth (24 percent) of the Japanese eighth-grade students achieved at or above the advanced benchmark.

Performance by students in the five top-performing East Asian countries far exceeded that of the other participants. Besides the East Asian countries, only in Hungary and the Netherlands did the share of students achieving the advanced benchmark reach double digits, 11 and 10 percent, respectively. In thirty countries, 5 percent of the students or less reached the advanced benchmark. In the United States, 7 percent of eighth-grade students reached the advanced benchmark, which was average for the participating countries.

Although Table 2-1 is organized to draw particular attention to the percentage of high-achieving students in each country, it also conveys information about the distribution of middle and low performers. Since students reaching the low benchmark may be considered to have gained some level of minimal competency in mathematics, it can be seen that the top-performing East Asian countries also do an excellent job in educating all of their students. Ninety-nine percent of the Singaporean eighth-graders reached the low benchmark as did 96 to 98 percent of the students in the other four top-performing East Asian Countries.

Trends in Mathematics Achievement

Table 2-2 shows, in alphabetical order, the countries that have comparable data from two or three TIMSS assessments. At the eighth-grade level, thirty-four countries had data from two assessments; seventeen of these had three-cycle trends. At the fourth grade, fifteen countries had comparable data from 1995 and 2003.

At the eighth grade, only five countries showed a pattern of significant improvement over the eight-year period from 1995 to 2003. One of these countries was the United States, which improved about 9 scale-score points from 1995 to 1999 and another 3 points from 1999 to 2003 for a significant 12-point increase over the eight years. Interestingly, two of the high-performing Asian countries also showed improvement over the eight-year period—Hong Kong SAR had a 17-point increase, and the Republic of Korea had an 8-point increase. The remaining two countries with improvement were two Baltic countries, including Latvia (LSS), with a 17-point increase for students in Latvian-speaking schools, and Lithuania, with a dramatic improvement of 30 scale-score points, 20 of which were from 1999 to 2003. The improvements in Latvia (LSS) and Lithuania, as well as the strong performance of Estonia in its first TIMSS appearance, indicate successful educational reforms in that region of the world.

Unfortunately, more countries declined significantly in mathematics achievement at the eighth grade than improved. Countries showing a decrease in TIMSS 2003 from 1995, 1999, or both, included Belgium (Flemish), Bulgaria, Cyprus, Iran, Japan, Macedonia, Norway, the Russian Federation, the Slovak Republic, Sweden, and Tunisia.

At the fourth grade, many countries had significant increases in average achievement between 1995 and 2003. Participants showing improved performance included Cyprus, England, Hong Kong SAR, Latvia (LSS), New Zealand, and Slovenia. In contrast, the Netherlands and Norway showed significant declines. For the United States, achievement at the fourth grade remained essentially the same—518 in both 1995 and 2003.

Contexts for Learning Mathematics

This section of the paper describes some of the most important and extensively studied factors related to improving mathematics achievement. At the early FIMS meetings in 1959, it was assumed that a cooperative approach to cross-national research would lead to better understanding of the more significant and effective educational procedures found anywhere in the world. In particular, the goal was to identify differences in education that were related to higher student achievement.

Even though a great deal of educational research is preoccupied with variations in instructional materials and strategies, the FIMS researchers observed that classrooms can look remarkably similar all over the world and concentrated on larger economic and societal issues where countries tended to have the largest differences. During the past forty years of conducting international mathematics assessments, however, much more has been learned about the conditions that can support and facilitate teaching and learning.

The TIMSS data show that the issues that matter the most are the fundamental ones of economic and cultural support for education, as well as the what, who, and how of teaching—a rigorous curriculum, well-prepared teachers, and effective instruction. There is little evidence that isolated factors consistently lead to improvement and excellence in mathematics achievement. Rather, it appears that high-achieving countries dedicate enormous effort and resources to ensuring systemic coherence across curriculum, teacher preparation, instruction, and assessment.

Economic Development

Since the national support for education is often related to level of economic development, figure 2-2 shows the relationship between average eighth-grade mathematics achievement and gross national per capita income for the coun-

Table 2-2a. *Trends in Average Mathematics Achievement—Eighth Grade*

Countries	2003 Average scale score	1999–2003 Difference		1995–2003 Difference	
Australia	505 (4.6)	–		–4 (6.0)	
Belgium (Flemish)	537 (2.8)	–21 (4.3)	▼	–13 (6.5)	▼
Bulgaria	476 (4.3)	–34 (7.3)	▼	–51 (7.2)	▼
Chile	387 (3.3)	–6 (5.5)		◇ ◇	
Chinese Taipei	585 (4.6)	0 (6.1)		◇ ◇	
Cyprus	459 (1.7)	–17 (2.4)	▼	–8 (2.8)	▼
Hong Kong, SAR	586 (3.3)	4 (5.4)		17 (7.0)	○
Hungary	529 (3.2)	–2 (4.9)		3 (4.5)	
Indonesia	411 (4.8)	8 (6.9)		◇ ◇	
Iran, Islamic Rep. of	411 (2.4)	–11 (4.1)	▼	–7 (4.5)	
Israel	496 (3.4)	29 (5.2)	○	–	
Italy	484 (3.2)	4 (5.0)		–	
Japan	570 (2.1)	–9 (2.7)	▼	–11 (2.6)	▼
Jordan	424 (4.1)	–3 (5.4)		◇ ◇	
Korea, Rep. of	589 (2.2)	2 (2.9)		8 (2.9)	○
Latvia (LSS)	505 (3.8)	0 (5.1)		17 (5.2)	○
Lithuania	502 (2.5)	20 (4.9)	○	30 (4.8)	○
Macedonia, Rep. of	435 (3.5)	–12 (5.5)	▼	◇ ◇	
Malaysia	508 (4.1)	–11 (6.0)		◇ ◇	
Moldova, Rep. of	460 (4.0)	–9 (5.6)		◇ ◇	
Netherlands	536 (3.8)	–4 (8.1)		7 (7.2)	
New Zealand	494 (5.3)	3 (7.4)		–7 (7.1)	
Norway	461 (2.5)	◇ ◇		–37 (3.3)	▼
Philippines	378 (5.2)	33 (7.9)h		◇ ◇	
Romania	475 (4.8)	3 (7.5)		2 (6.6)	
Russian Federation	508 (3.7)	–18 (7.0)i		–16 (6.5)	▼
Scotland	498 (3.7)	◇ ◇		4 (6.8)	
Singapore	605 (3.6)	1 (7.2)		–3 (5.4)	
Slovak Republic	508 (3.3)	–26 (5.2)	▼	–26 (4.5)	▼
Slovenia	493 (2.2)	–		–2 (3.7)	
South Africa	264 (5.5)	–11 (8.8)		–	
Sweden	499 (2.6)	◇ ◇		–41 (5.0)	▼
Tunisia	410 (2.2)	–38 (3.3)	▼	◇ ◇	
United States	504 (3.3)	3 (5.2)		12 (5.8)	○
TIMSS scale avg.	500				

tries participating in TIMSS 2003. It is noteworthy that the wealthiest countries (those with gross national per capita income of $20,000 or more) are found in the top half of the achievement distribution, while many of the poorer countries are in the lower half. The pattern is by no means clear-cut, however. For example, Korea and Chinese Taipei, two of the highest-performing coun-

Table 2-2b. *Trends in Average Mathematics Achievement —Fourth Grade*

Countries	2003 Average scale score	1995–2003 Difference	
Australia	499 (3.9)	4 (5.2)	
Cyprus	510 (2.4)	35 (4.0)	◐
England	531 (3.7)	47 (5.0)	◐
Hong Kong, SAR	575 (3.2)	18 (5.1)	◐
Hungary	529 (3.1)	7 (4.8)	
Iran, Islamic Rep. of	389 (4.2)	2 (6.5)	
Japan	565 (1.6)	−3 (2.4)	
Latvia (LSS)	533 (3.1)	34 (5.5)	◐
Netherlands	540 (2.1)	−9 (3.6)	▼
New Zealand	496 (2.1)	26 (4.9)	◐
Norway	451 (2.3)	−25 (3.8)	▼
Scotland	490 (3.3)	−3 (5.3)	
Singapore	594 (5.6)	4 (7.2)	
Slovenia	479 (2.6)	17 (4.1)	◐
United States	518 (2.4)	0 (3.8)	
TIMSS scale avg.	500		

Source: IEA's Trends in International Mathematics and Science Study (TIMSS) 2003.

Trend notes (Eighth Grade): Because of differences in population coverage, 1999 data are not shown for Australia and Slovenia, and 1995 data are not shown for Israel, Italy, and South Africa. Korea tested later in 2003 than in 1999 and 1995, at the beginning of the next school year. Similarly, Lithuania tested later in 1999 than in 2003 and 1995. Data for Latvia in this exhibit include Latvian-speaking schools only.

Trend notes (Fourth Grade): Because of differences between 1995 and 2003 in population coverage, 1995 data are not shown for Italy. Data for Latvia in this exhibit include Latvian-speaking schools only. To be comparable with 1995, 2003 data for New Zealand in this exhibit include students in English medium instruction only (98% of the estimated population).

( ) Standard errors appear in parentheses. Because results are rounded to the nearest whole number, some totals may appear inconsistent.

A dash (–) indicates comparable data are not available.

A diamond (✧) indicates the country did not participate in the assessment.

◐ = 2003 Country average significantly higher.

▼ = 2003 Country average significantly lower.

tries, have modest national per capita income. Also, many of the former Eastern bloc countries have relatively low per capita income but relatively high average mathematics performance, perhaps the result of a traditional emphasis on mathematics education.

The Organization of Schools

The international studies of mathematics achievement have demonstrated that the school structure itself is one of the most important variables accounting for differences in student achievement. The types of schools that exist in a country,

Figure 2-2. *TIMSS 2003 Eighth Grade Mathematics Achievement by Gross National Income per Capita*

Eighth grade mathematics achievement

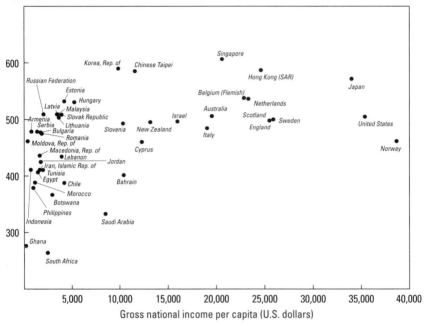

Gross national income per capita (U.S. dollars)

Source: World Bank (2004).

how children progress through the schools, and how many schools are found at each stage all matter. Whether universal education exists for all students through the secondary level is also an important variable. Although many of the world's nations still have major challenges in implementing universal education, considerable progress has been made in this area since FIMS. Of the twelve original participating countries, nine were European and the other three were Australia, Japan, and the United States—all high-income, technologically developed nations that realized the importance of raising the upper age of compulsory education and providing equality of educational opportunity for all. Still, among this group of countries the percentage of 16-year-olds remaining in school in the early 1960s ranged from less than 25 percent in England to about 86 percent in the United States.[18] The FIMS choice of the grade with the most 13-year-olds as a target population was based on the fact that students in some of the FIMS countries began leaving school after age 13. One of the most significant developments in education systems around the world in the years since the first international mathematics assessment has been the large increase in the number of students completing upper secondary education. Unfortunately, considerable variation in completion rates still exists among countries.

Besides the universality of education, there remains the issue of differentiation or specialization, usually depending essentially on whether students are bound for university. Each country in the world has a set of policies and practices that influences the allocation and selection of students to each type of secondary school program and that also determines the proportion of students who will complete each program, transfer to another program, or drop out of school.

This issue of completion rates as well as the percentages of students having taken advanced mathematics and physics was studied most recently in TIMSS 1995 as part of an assessment of the school-leaving population. To learn how much of the school-leaving cohort was still in school and represented by the TIMSS sample, a TIMSS Coverage Index (TCI) was computed for each country. The TCI was computed by forming a ratio of the size of the student population covered by the TIMSS sample, as estimated from the sample itself, to the size of the school-leaving age cohort, which was derived from official population census figures supplied by each country.[19] (The school-leaving age cohort is the population that is the age that should be graduating in any given year.) One component of the school-leaving age cohort not covered by the TIMSS sample was the percentage of students no longer attending school. This percentage varied greatly, from more than 50 percent in South Africa to about 35 percent in the United States to 12–15 percent in Norway and France. Another component of the school-leaving age cohort is the percentage of students who have taken advanced courses. The percentage of students who have taken advanced mathematics ranged from 2 percent in the Russian Federation to 75 percent in Slovenia. For physics, the percentages ranged from 3 percent in the Russian Federation to 44 percent in Austria and Slovenia. One important aspect of TIMSS Advanced 2008 will be estimating any changes in the size and achievement of the population of students taking advanced courses in mathematics and physics.

Curriculum

Much has been learned about the impact of the curriculum on mathematics achievement. This factor was first looked at with SIMS and then studied extensively in TIMSS 1995.[20] Since the mathematics curricula across countries are discussed in particular elsewhere in this volume, the discussion here is confined to an overview of the most recent findings in TIMSS 2003.

In TIMSS 2003 most countries reported having the same intended curriculum for all students, with no grouping of students. At the eighth-grade level, thirty-eight participants reported this approach, while nine countries reported having one curriculum for all students, but at different difficulty levels for groups of students with different ability levels. Only four countries—Belgium (Flemish), the Netherlands, the Russian Federation, and Singapore—reported having different curricula for different groups of students according to their ability level. At the fourth-grade level, all participants reported having just one curriculum for all students; in most cases there was no grouping by ability level.

Although the relationship between inclusion in the intended curriculum and student achievement was not perfect in TIMSS 2003, it was notable that several of the higher-performing countries reported high levels of emphasis on the mathematics topics in the intended curricula. For example, five of the six top-performing countries at the eighth grade (Chinese Taipei being the exception) included 80 percent or more of the topics in the intended curriculum for all or almost all of their students. On average, 70 percent of the topics were in the intended curriculum for all or almost all students, and, for countries with differentiation by ability, another 6 percent of the topics were in the intended curriculum for students in the top academic track. In only six countries were fewer than half of the topics included, and as might be anticipated, these countries had relatively low achievement (Botswana, Indonesia, Lebanon, Morocco, the Philippines, and Tunisia). Similarly, at the fourth-grade level, higher-performing countries had generally greater levels of coverage (for example, about 70 percent for Japan and Singapore); and lower-performing countries, lesser levels (about 20 percent for Morocco and Tunisia).

In the recent past, it became common for countries' intended curricula to be updated regularly. At the time of the TIMSS 2003 testing, the official eighth-grade mathematics curriculum was undergoing revision in about three-fourths of the countries and had been in place for five years or less in twenty-seven countries. At the fourth-grade level, about half the countries were undergoing curriculum revision and twenty had a curriculum that had been in place for five years or less.

Having a rigorous curriculum is fundamental to exposing students to increasingly challenging mathematical concepts and ideas; however, it is up to the teacher to actually provide classroom instruction in the topics. Given that the curriculum may be becoming somewhat of a moving target, it is interesting to note that teachers at both grades said 72–73 percent of their students, on average, had been taught the TIMSS 2003 mathematics topics. At the eighth grade, there was substantial agreement between the topics in the intended curriculum and the topics taught, but this was much less so at the fourth grade.

Teacher Experience and Preparation

Since the teacher can be a powerful force in determining the quality and quantity of educational achievement, each international study of mathematics achievement beginning with FIMS has sought to relate teacher variables to the test results. Among the variables are the teacher's level of training, amount and quality of teaching experience, and status as a professional worker. While the data show little evidence of a direct relationship between teacher training and student achievement, the studies have collected considerable information about policies and practices across countries.

For example, TIMSS 2003 collected information about five requirements for being certified or licensed to teach mathematics at the fourth and eighth grades, including obtaining a university degree, supervised practical experience (practicum), passing an examination, completion of a probationary period, and completion of an induction program. Most countries required only two or three of these. Taking the requirements one by one for the forty-seven participating countries at the eighth grade, thirty-three of the countries required a university degree (or equivalent), thirty-three required some type of practicum, twenty-eight required passing an examination, and twenty-three a probationary period. At the fourth grade, nineteen of twenty-six countries required some type of practicum, and eighteen required two or more of the following—passing an examination, a university degree, or completion of a probationary period. Only eleven countries at eighth grade and eight at fourth grade required completion of an induction program.

Teachers' reports about their preparation to teach mathematics revealed that approximately three-fourths of eighth-grade students and two-thirds of fourth-grade students were taught mathematics by teachers having at least a university degree or equivalent. At the eighth grade, on average, the majority of students had teachers who had studied mathematics (70 percent) or mathematics education (54 percent) or both (since teachers often reported that their study was focused in more than one area). As might be anticipated, the situation was different at the fourth grade, where teachers typically studied primary or elementary education (for 80 percent of the students, on average).

Despite a certain lack of uniformity in their preparation and training, and achievement from students indicating the contrary, the teachers at both grades reported that they felt ready to teach nearly all the TIMSS 2003 topics. At the eighth-grade level, at least 90 percent of the students had teachers who felt ready to teach all of the topics, with the exception of three of the data topics (sources of error, data collection methods, and simple probability). At the fourth-grade level, the results dipped below 90 percent for only two geometry topics (relationships between two- and three-dimensional shapes; and translation, reflection, and rotation).

Instruction

Since the types of learning experiences provided for students can vary from country to country, a significant area of inquiry in each international study of mathematics achievement has been whether different experiences lead to different achievement levels. One obvious area of inquiry has been the amount of mathematics instruction given to students, since research about time on task and common sense suggest this as a prerequisite to learning. The TIMSS results, however, are not as conclusive as might be anticipated. For example, in TIMSS 2003, countries that reported providing the most instruction (more than 150 hours per year) at the eighth-grade level had relatively low achievement (the Philippines, Indonesia, and Chile). It is apparent that time alone is not enough and that the time provided must be used effectively.

The instructional strategies used in mathematics classrooms around the world are remarkably similar. TIMSS consistently has found the textbook to be the foundation of instruction. In 2003 nearly two-thirds of the fourth- and eighth-grade students had teachers who reported using a textbook as the basis of their lessons. The three predominant activities at both grades, accounting for about 60 percent of class time, were teacher lecture, teacher-guided student practice, and students working on problems on their own.

Using technology to facilitate mathematics teaching and learning has emerged as a topic of considerable study beginning with TIMSS 1995. Most recently, TIMSS 2003 found that calculator use at the eighth grade varied dramatically from country to country. The countries permitting calculator use for nearly all students (98 percent or more) included both high-performing Hong Kong SAR and low-performing Morocco. At the fourth grade, 57 percent of the students, on average, were not permitted to use calculators. The countries permitting widespread calculator usage (90 percent of the students or more) all performed at about the international average—Australia, Cyprus, England, New Zealand, and Scotland. Even in those countries, however, teachers reported asking relatively small percentages of students to do any calculator activities in half the lessons or more. Access to computers remains a challenge in many countries. Internationally, teachers reported that, on average, computers were not available for 68 percent of the eighth-grade students and 58 percent of the fourth-grade students. Even in countries with relatively high availability, using computers as often as in half the lessons was extremely rare (typically for 5 percent of the students or less) at either grade for any purpose. The greatest use at fourth grade was by the Netherlands and Singapore, where 31 and 14 percent of the students, respectively, used the computer for practicing skills and procedures.

Home Support for Learning

IEA's ongoing international assessments have shown that in almost every country students from homes with extensive educational resources have higher achievement in mathematics than those from less advantaged backgrounds.

Most recently, TIMSS has focused on just a few central variables: level of parental education, students' educational aspirations, speaking the language of the test at home, having a range of study aids in the home, and computer use at home and at school.

TIMSS 2003 data make it clear that higher levels of parents' education are associated with higher eighth-grade student achievement in mathematics in almost all countries. Further, students with university-educated parents are particularly likely to expect to attend university themselves.

At both eighth and fourth grades, TIMSS 2003 found that students from homes where the language of the test is always or almost always spoken had higher average achievement than those who spoke the test language less frequently. Even though many countries tested in more than one language in order to cover their whole student population, on average, about 21 percent of students were from homes where the language of the test was spoken only sometimes or never. Countries where the majority of students were from homes where the language of the test was spoken infrequently generally had relatively low mathematics achievement, including Botswana, Ghana, Indonesia, Lebanon, Morocco, the Philippines, and South Africa.

TIMSS 2003 (and previous studies) also found a clear-cut relationship, on average, between the number of books in the home and mathematics achievement at both grades. In addition to literacy resources, having study aids such as a computer or a study desk or table at home was associated with higher student achievement. Mathematics achievement was positively related to computer usage, particularly at eighth grade, with average achievement highest among students reporting using computers at home and at school. At both grades, the data indicated that students were somewhat more likely to use a computer at home than at school. It should be emphasized, however, that at both grades the percentages of students reporting that they did not use a computer at all varied dramatically across countries. For example, the percentages of eighth-grade students not using a computer ranged from 0 percent in Hong Kong SAR and Chinese Taipei to about two-thirds in Botswana and Iran.

Attitudes and Values

The FIMS researchers noted that even though attitudes and values are directly related to educational practices and outcomes, this is a difficult area to measure, and the inferences about cause and effect are more speculative. Nevertheless, the researchers developed and administered six different scales about teaching and learning mathematics, the role of mathematics in society, and attitudes toward the environment to secure data on attitudes and values for descriptive purposes and later study.

More than four decades later, TIMSS 2003 reported results for two attitudinal scales—one about students' self-confidence in learning mathematics, and one on the value students place on mathematics. At both grades, students had higher mathematics achievement, on average, if they were more self-confident in learning mathematics and placed higher value on the subject. It is clear that students' motivation to learn mathematics can be affected by whether they find the subject enjoyable, place value on the subject, and think it is important for success in school and for future career aspirations. Nevertheless, in some countries with high average mathematics achievement (Korea, Japan, and the Netherlands), less than 20 percent of their students placed a high value on learning mathematics. Some research has indicated that the students following a demanding curriculum may have higher achievement but little enthusiasm for the subject matter.[21]

Safe and Orderly Schools

Unfortunately, one of the emerging factors for study in TIMSS is school safety. TIMSS 2003 asked both teachers and students to characterize their perceptions of safety in their schools. At both grades, there was a positive relationship between teachers' reports of school safety and mathematics achievement. About three-fourths of the students were in schools the teachers characterized as safe; however, one-fifth were in schools where teachers were only in partial agreement on safety, and about 4 to 6 percent were in schools deemed not to be safe. The students were asked about having things stolen, being hit or hurt, being made to do things, being made fun of, and being left out of activities. Fifteen percent of the eighth-grade students and 23 percent of the fourth-grade students answered "yes" to all five statements, and these students had lower average mathematics achievement than their counterparts in better school environments.

Summary

The FIMS pioneers of international mathematics assessment were experts in sampling, test and questionnaire construction, and statistical data analysis that sought to expand their field of expertise across countries to gain further insight into effective educational practices. By conducting an international assessment involving twelve economically developed countries, they sparked the possibilities for the future. Keeping in mind the model of the intended, implemented, and achieved curriculum provided by SIMS, as well as the rigor of comparative validity and innovative methodological advances introduced by TIMSS, many lessons have been learned over the past decades about how to conduct international assessments of high quality.

Much also has been learned about the educational process, and about what matters in teaching and learning mathematics. Education is an arduous process. Different countries use different approaches, but effective education always requires enormous effort. Success requires developing a rigorous and progressive curriculum and providing all students with an equal opportunity to learn it. Success also depends on economic resources and a strong-willed society to ensure that students are ready to learn and that teachers are well prepared to provide instruction, as well as having the necessary facilities and materials.

TIMSS 2007 data collection has begun, with more than sixty countries expected to take part by June 2007. TIMSS Advanced, for students in the final year of secondary school (twelfth grade in most countries), will occur in 2008 and planning is under way for TIMSS 2011. With the explosion of technological discoveries facilitating improved measurement techniques as well as international communications, and the increasing number of diverse countries striving to become full participating members of our global enterprise, it seems that international assessments will continue to grow in both the quality of the data provided and the quantity of the participating countries.

Notes

1. Torsten Husén, ed., *International Study of Achievement in Mathematics,* 2 vols. (New York: John Wiley, 1967).
2. Ibid.
3. Analyses of the intended and implemented curricula were reported in Kenneth J. Travers and Ian Westbury, eds., *The IEA Study of Mathematics I: Analysis of Mathematics and Curricula* (Oxford, England: Pergamon Press, 1989).
4. David F. Robitaille and Robert A. Garden, eds., *The IEA Study of Mathematics II: Contexts and Outcomes of School Mathematics* (Oxford, England: Pergamon Press, 1989).
5. William H. Schmidt and others, *Many Visions, Many Aims,* vol. 1: *A Cross-National Investigation of Curricular Intentions in School Mathematics* (Dordrecht, The Netherlands: Kluwer Academic Publishers, 1997).
6. James W. Stigler and others, *The TIMSS Videotape Classroom Study: Methods and Findings from an Exploratory Research Project on Eighth-Grade Mathematics Instruction in Germany, Japan, and the United States,* NCES 1999-074 (U.S. Department of Education, National Center for Education Statistics, 1999).
7. Harold W. Stevenson and Roberta Nerison-Low, *To Sum It Up: Case Studies of Education in Germany, Japan, and the United States* (U.S. Department of Education, Office of Educational Research and Improvement, National Institute on Student Achievement, Curriculum, and Assessment, 1999).
8. David F. Robitaille and others, *Curriculum Frameworks for Mathematics and Science* (Vancouver: Pacific Educational Fund, 1993).
9. Maryellen Harmon and others, *Performance Assessment in IEA's Third International Mathematics and Science Study* (Boston College, 1997).
10. For descriptions, see Albert E. Beaton and others, *Mathematics Achievement in the Middle School Years: IEA's Third International Mathematics and Science Study* (Boston College, 1996); Ina V. S. Mullis and others, *Mathematics Achievement in the Primary School Years: IEA's Third International Mathematics and Science Study* (Boston College, 1997); and Ina V. S. Mullis

and others, *Mathematics Achievement in the Final Year of Secondary School: IEA's Third International Mathematics and Science Study* (Boston College, 1998). The technical and operational aspects of TIMSS, and details of the methodology employed, are fully described in a series of technical reports; see Michael O. Martin and Dana L. Kelly, eds., *Third International Mathematics and Science Study Report*, vol. 1: *Design and Development* (Boston College, 1996); vol. 2, *Implementation and Analysis, Primary and Middle School Years* (Boston College, 1997); vol. 3, *Implementation and Analysis* (Boston College, 1998).

11. See, for example, David F. Robitaille, Albert E. Beaton, and T. Plomp, eds., *The Impact of TIMSS on the Teaching and Learning of Mathematics and Science* (Vancouver: Pacific Educational Press, 2000).

12. See, for example, Michael O. Martin and Ina V. S. Mullis, eds., *Third International Mathematics and Science Study: Quality Assurance in Data Collection* (Boston College, 1996).

13. Ina V. S. Mullis and others, *TIMSS 1999 International Mathematics Report: Findings from IEA's Trends in International Mathematics and Science Study at the Eighth Grade* (Boston College, 2000).

14. Ina V. S. Mullis and others, *TIMSS 2003 International Mathematics Report: Findings from IEA's Trends in International Mathematics and Science Study at the Fourth and Eighth Grades* (Boston College, 2004).

15. Ina V. S. Mullis and others, *TIMSS 2007 Assessment Frameworks* (Boston College, 2005).

16. Robert A. Garden and others, *TIMSS Advanced Assessment Frameworks* (Boston College, 2006).

17. For a full set of findings and a description of procedures, please see Mullis and others, *TIMSS 2003 International Mathematics Report.*

18. Husén, *International Study of Achievement in Mathematics.*

19. Mullis and others, *Mathematics Achievement in the Final Year of Secondary School.*

20. For the SIMS study, see Travers and Westbury, *The IEA Study of Mathematics I.* For the TIMSS study, see Schmidt and others, *Many Visions, Many Aims.*

21. Ce Shen, "Revisiting the Relationship between Students' Achievement and Their Self-Perceptions: A Cross-National Analysis Based on TIMSS 1999 Data," *Assessment in Education* 9 (2): 161–84.

3

Understanding Causal Influences on Educational Achievement through Analysis of Differences over Time within Countries

JAN-ERIC GUSTAFSSON

The International Association for the Evaluation of Educational Achievement (IEA) was founded in 1959 by a small group of education and
social science researchers with the purpose of using international comparative
research to understand the great complexity of factors influencing student
achievement in different subject fields. A popular metaphor was that they
wanted to use the world as an educational laboratory.

The first study, which investigated mathematics achievement in twelve
countries, was conducted in 1964.[1] Since the publication of that study, different groups of researchers have published, under the auspices of the IEA, a large
number of studies of education achievement in different countries in a wide
range of subject areas. For example, in the first round in 1995 of the TIMSS
study (at that time TIMSS was an acronym for Third International Mathematics and Science Study), which investigated knowledge and skill in mathematics
and science, forty-five countries participated.[2] For the third round of TIMSS
(TIMSS now stands for Trends in International Mathematics and Science
Study),[3] which was conducted in 2003, more than fifty countries participated,
and for the fourth round in 2007 an even larger number of countries will participate. Not only has the number of participating countries increased dramatically, but so too has the frequency of repetition; the studies of mathematics,

science, and reading are now designed to capture within-country achievement trends and are therefore repeated every fourth or fifth year.

The data collected in these studies have been used to generate a vast amount of knowledge about differences in achievement from country to country. Because the data have been made freely and easily available, they have been used in a large number of secondary analyses by researchers in different fields such as economics, education, sociology, and didactics of different subject matter areas. Voices of criticism also have been raised against them, however.

One line of criticism holds that the international studies have come to serve primarily as a source of benchmarking data for purposes of educational policy and debate, thereby becoming a means of educational governance that reduces the importance and influence of national policymakers.[4] This benchmarking function grew in importance during the 1990s, in part because the international studies adopted the methodology developed in the National Assessment of Educational Progress (NAEP) in the United States in the 1980s, which was based on complex item-response theory and matrix-sampling designs.[5] This methodology was well suited for making efficient and unbiased estimations of country-level performance. In addition, the increasing number of participating countries made the benchmarking function more interesting. That became even more pronounced when the Organization for Economic Cooperation and Development (OECD) also started international surveys of educational achievement through the Program for International Student Assessment (PISA).[6] The OECD presence led to even greater emphasis on the economic importance of the educational results.

Another line of criticism contends that the "world educational laboratory" has not been particularly successful in disentangling the complex web of factors that produce a high level of knowledge and skills among students. Even though advances have been made, a great deal remains to be learned, and doubts have been expressed that cross-sectional surveys offer the appropriate methodology for advancing this kind of knowledge. Indeed, Allardt argues that there is little evidence that comparative surveys in any field of social science have been able to generate knowledge about causal relations.[7] He points to the great complexity of the phenomena investigated, and to the uniqueness of different countries, as the reasons for this. The observation that cross-sectional surveys do not easily allow causal inference is made in many textbooks on research design, so the methodological challenges are well known. Furthermore, the international studies of educational achievement are not based upon an elaborated theoretical framework, which makes it difficult to apply the analytical methods developed for making causal inference from cross-sectional data.[8]

It would be unfortunate if these difficulties prevented researchers from seeking the causes behind the patterns revealed in comparative educational surveys.

The search for explanations is one of the main aims of scientific research, and explanations also are needed if policymakers are to take full advantage of the benchmarking results. This chapter argues that causal inferences might be more robust if researchers took advantage of the possibilities for longitudinal analyses at the country level of the data generated by the trend design implemented in the latest generation of the comparative educational studies conducted by the IEA and OECD. With the longitudinal design many variables that cause spurious results in cross-sectional research are kept constant because they do not vary within countries over time.

The chapter thus has two main aims. First, it discusses those methodological problems and pitfalls in the international studies that are of importance when the purpose is to make causal inferences. Second, it uses two concrete examples to show how a longitudinal approach to analysis of country-level trend data can avoid some of these methodological problems.

Problems and Pitfalls in Causal Inference from Cross-Sectional Data

A typical international cross-sectional study measures achievement for samples of students in a set of participating countries. Information also is collected about student characteristics (such as social background, gender, and courses taken), teacher characteristics (experience and education), teaching characteristics (teaching methods, homework), and school characteristics (size, location, and relations with parents). In addition, country-level information is typically collected about, among other things, curriculum and institutional factors (such as organization of the school system and decision-making power at different levels). Using different kinds of statistical analysis, such as regression analysis, these background and contextual factors are treated as independent variables to determine the amount of their influence on achievement.

Let me take a concrete example of such a study. A study reported by Mullis, Campbell, and Farstrup used NAEP data collected in 1992 to investigate the effects on reading performance of the amount of direct teaching of reading received by fourth-grade students.[9] When the researchers correlated the amount of instruction with achievement, they found a significant negative correlation—the students who had obtained more teaching had a lower level of reading achievement. It does not seem reasonable, however, to interpret the negative correlation to mean that more direct teaching of reading causes students to read more poorly. A more reasonable explanation for the negative correlation is that direct teaching of reading was part of a compensatory educational strategy, in which poor readers were provided with more teaching resources, either in regular education or in special education, than were proficient read-ers. The authors of the report favored this interpretation.

This example suggests that it is easy to confuse the direction of causality in cross-sectional data, and that caution should be exercised when one makes causal statements based on analyses of such data. The problem of confused direction of causality is well known in virtually every discipline. Econometri-cians refer to it as the "endogeneity" problem, while sociologists and psychol-ogists often talk about the problem of "reversed causality." Another term for it is "selection bias," that is, the levels of performance of groups of students who received different treatments were not comparable before they received the treatments, which biased the results of the treatment.

This problem, whatever label is used, seems inescapable in studies with a cross-sectional design, at least when the analysis is done at the individual level. Rarely, if ever, is it reasonable to assume that the amount and kind of instruction allo-cated to a student is independent of the characteristics of the student. As the example given earlier illustrates, compensatory resource allocation is common in educational contexts, implying that more or better instruction is given to poorly performing students than to better-performing students. Another example comes from Lazear, who has developed a theoretical model to account for the effects of variation in class size. The model explains, among other things, why there will be a selection bias such that larger classes tend to be populated by higher-performing students.[10] Thus, a positive correlation may be expected between class size and achievement, which, however, only reflects the selection bias.

One way to deal with the problem of selection bias is to control statistically for the differences between students that existed before the treatment was applied. This approach requires a measure of these preexisting differences, however, and in a cross-sectional design such measures are generally not avail-able. If it is possible to use a longitudinal design, one measure may be made before the treatment is applied, and one or more measures can be made after the treatment. Longitudinal designs have not been used in international studies of educational achievement, mainly for practical reasons. Instead other, more readily available measures, such as indicators of socioeconomic background, have been used to control for the preexisting differences.

This brings me to another source of erroneous causal inference from cross-sectional data, namely, the problem of omitted variables. When an indepen-dent variable is related to a dependent variable in a statistical model and the estimated relation is interpreted in causal terms, it is assumed that there are no other independent variables that are correlated with the independent variable being studied and that have not already been included in the model. If such omitted variables do exist, they will cause bias in the estimated causal relations if they correlate with the residual of the dependent variable, possibly leading researchers to ascribe causality to variables other than the ones that are actually involved. One approach to solving the problem of omitted variables would be to measure and analyze all potentially relevant variables. It is, however, virtually impossible to include all the relevant variables, even if strong theory is available to guide the selection.

To summarize the discussion so far, two main threats to correct causal inference from cross-sectional data have been identified. The first is the problem of selection bias, which not only confuses the direction of causality but also causes biased estimates of the strength of effects. The other problem is that of omitted variables, which makes us ascribe causality to independent variables other than the ones that are actually causally involved.

Wößmann's Secondary Analysis of TIMSS 1995

Before proceeding, it may be useful to look at a concrete example of an analysis of IEA data that tries to solve the problems of causal inference, namely, a study by Ludger Wößmann. Wößmann recently presented a very ambitious and interesting analysis of the TIMSS 1995 data for grades seven and eight, otherwise known as Population B.[11] The main goal of the study was to investigate effects of schooling resources and institutional factors on student achievement.

Wößmann argued that the data should be analyzed at the individual level:

> The relevant level of estimation is the individual student (not the class, school, district or country), because this directly links a student's performance to the specific teaching environment. The estimation of such microeconometric education production functions provides the opportunity to control for individual background influences on student performance, to assess the effects of the relevant resource and teacher characteristics which the student faces, and to estimate the effects of institutional features below the country level which are relevant to the individual student.[12]

He thus constructed a student-level database comprising the more than 260,000 students in grades seven and eight from the thirty-nine countries that participated in TIMSS 1995. The database included the achievement scores in mathematics and science; information collected through questionnaires administered to students, teachers, and principals; and country-level information collected in the TIMSS study that had been combined with data from other sources on the institutional structure of the educational system, such as the level of decisionmaking and use of centralized examinations.

The data were analyzed with regression models, in which student test scores were regressed on measures of student background, along with indicators of resources and institutional factors measured at the classroom, school, and country level. To take into account effects of the hierarchically structured data on estimates of standard errors, Wößmann used a technique called clustering-robust linear regression, which empirically estimates the covariances of the error terms caused by the nonindependence of individuals attributable to the cluster sampling design.

Wößmann expressed confidence that this model would correctly estimate effects of the country-level institutional factors, because there should not be any endogeneity problems in the estimation of these effects. He also emphasized that estimation of these factors requires using the world as an educational laboratory: "The link between institutions and student performance could hardly be tested using country-specific data as there is no significant variation in many institutional features within a single country on which such an analysis could be based. Only the international evidence which encompasses many education systems with widely differing institutional structures has the potential to show whether institutions have important consequences for students."[13]

Wößmann's main conclusion was that institutional factors had very strong effects on student performance. Among the positive effects were the performance of centralized examinations and control mechanisms, school autonomy in making decisions about hiring of teachers, and individual teacher influence over teaching methods. The results also showed that resource factors such as class size had little effect on student achievement. Wößmann noted, however, that "given the single cross-section structure of the performance study, the potential endogeneity problems plaguing resource estimates cannot be fully overcome in the TIMSS data."[14] Thus, the negative findings concerning effects of resource factors should be interpreted with caution.

It may be instructive to take a closer look at some of the estimates obtained to judge whether they are reasonable. For example, Wößmann found that attending an additional grade improved mathematics performance by 40 points on the scale with a mean of 500 and a standard deviation of 100. This estimate is in line with, although higher than, most other studies of the effects of attending another year of school.[15] A negative effect of -14 points was observed, however, for becoming one year older in chronological age. This estimate is hard to accept as a correct determination of the effect of becoming one year older, because it is a well-established finding that chronological age is positively associated with achievement in mathematics in the middle school years. Instead it is more reasonable to interpret the negative effect of age as being due to endogeneity problems: students of lower ability start school at an older age and repeat classes more often than high-performing students. Obviously the available measures of student characteristics (primarily family background) did not adequately control for the selection bias introduced in this way.

For class size a fairly strong and significant positive effect was found, implying that students who attended larger classes outperformed students who attended smaller classes. This is counterintuitive and does not agree with findings in a large number of other studies (see below). For this age level the most well-established finding is that class size does not have any effect on achievement.[16] Again, the most credible interpretation of this effect is that it is an expression of an endogeneity problem, which cannot be properly controlled for because of a lack of entry-level measures of achievement in the TIMSS data.[17]

The interpretation of the findings concerning institutional effects also raises some issues. One concern is that omitted variables may be causing estimates of parameters to be biased, and incorrect inferences to be made about which independent variable is causing the effect. For example, one of Wößmann's main findings was that use of centralized examinations has an effect on achievement (16 points for mathematics). However, the use or nonuse of centralized examinations has not been randomly assigned to countries; their use tends to be more common in certain categories of countries (East Asian countries in particular) than in other categories of countries. The different categories of countries also are likely to differ in a large number of other respects, such as the value ascribed to education and how teaching in classrooms is organized. Such variables may also affect achievement, and if they are not controlled for, the effect may be incorrectly ascribed to use of centralized examinations. Given the great number of potential omitted variables, it is an impossible task to argue that a particular study is not afflicted by the omitted-variable problem.

Alternative Approaches to Causal Inference

It thus seems that the two major methodological problems of endogeneity and omitted variables still remain to be solved in the analysis of data from international studies of educational achievement. Some alternative approaches for dealing with these problems are offered here.

The lack of appropriate statistical control over previous achievement makes it very difficult to deal with the problem of selection bias in the comparative studies of educational achievement. Even though selection bias seems virtually impossible to deal with at the individual level, it may be more approachable at higher levels of aggregation. Thus, even though there may inextricable connections between student characteristics and the treatments that the students receive, such connections need not exist at the school or national level. A concrete example of this is the suggestion by Wößmann and West to study effects of class size at the school level rather than the class level.[18] The distribution of students over classes within a school is likely to be associated with endogeneity problems, but there is no a priori reason to suspect that mean class size for the schools is affected by endogeneity problems.

Another example is the amount of homework assigned to students. At the student level selection bias may occur in that poorly performing students are assigned much homework, which at least some students spend quite a lot of time on. Motivated and high-achieving students also are likely to spend time and effort on homework. The combined effect may be little relation between amount of homework and student achievement. However, teachers also may vary in their readiness to assign homework to students, which is an expression of an instructional strategy and which is independent of the composition of the group of students. One way to conduct such analyses would be to perform two-level modeling of the relations between amount of homework and achievement at the individual level and at the class level using either regression techniques as implemented in hierarchical linear models, or two-level latent variable modelling techniques as implemented in, for example, the Mplus program.[19]

The analysis could also be conducted at even higher levels of aggregation, such as the national level. At this level, the risk of being misled by problems of selection bias will be even lower, because it is not reasonable to expect that differences in use of homework, for example, between countries can be affected by endogeneity problems.

Hanushek and Wößmann have conducted a very interesting study using data aggregated to the country level.[20] They analyzed effects of educational tracking on level and inequality of performance through investigating differences between higher grades and lower grades within countries with respect to dispersion and mean level of performance. In this way they controlled for between-country differences, and the observed difference could be related to whether educational tracking was used or not. The results showed that educational tracking increases inequality of educational achievement and also tends to depress level of performance.

Analysis of data at higher levels of aggregation may thus be one way of solving the problem of selection bias. There may be reason to use different levels of aggregation for different substantive questions, but here I assume aggregation to the country level. The main reason for this choice is that several of the international studies now are designed to yield estimates of change in achievement over time. Such trend data on a panel of countries offer opportunities for longitudinal analysis, which may help solve at least a part of the omitted variable problem. Through relating within-country change over time in explanatory variables to within-country change in achievement, a "fixed-country" analysis is performed that keeps most country characteristics constant. These fixed characteristics cannot be considered as omitted variables.

While longitudinal analyses of data aggregated to the country level allow investigation of only a subset of questions of interest in comparative educational research, it may be worthwhile to see to what extent causal inference is possible.

Two examples of research issues are discussed, mainly from the methodological point of view, namely, the effect of student age on achievement, and the effect of class size on achievement.

Example 1: Student Age and Mathematics Achievement

The analytical idea is illustrated with a concrete example, focusing on an independent variable that should have a fairly simple relationship with mathematics achievement. One such variable is student age, which is positively related to student achievement according to results from several different kinds of studies.

The TIMSS 1995 study allowed inference about the combined effect of another year of schooling and becoming one year older, because adjacent grades were sampled for population 1 (9–10-year-olds) and population 2 (13–14-year-olds). For population 1 most countries sampled grades three and four, and the international mean in performance for grade four was 57 points higher than for grade three. The increases in mean performance between the two grades varied between a high of 84 points in the Netherlands and a low of 46 points in Thailand. For population 2 the samples of most countries included students from grades seven and eight, and the international mean difference in level of performance between upper and lower grades amounted to 30 points, with a high mean difference of 47 points in Lithuania and 7 points in South Africa.

These results indicate that the combined mean age-grade effect is almost twice as large for the younger students as it is for the older students. As the TIMSS report noted, however, the differences in level of achievement in grades seven and eight were affected in some cases by policies regarding student promotion from one grade to another, which caused the performance difference to be underestimated.[21] There may thus be some downward bias in the between-grade estimates, but the results for both populations nevertheless indicate a sizable age effect.

These estimates also agree quite well with results presented by Cahan and Cohen, who furthermore showed that with a regression discontinuity design the effects on achievement of one year of schooling could be separated from the effects of becoming one chronological year older.[22] Chronological age accounted for about one-third of the total effect, while schooling accounted for the remaining two-thirds.

Results for Grade Eight

To estimate the age-grade effect from trend data for grade eight, a database has been constructed using information from twenty-two countries that participated in TIMSS 1995 and TIMSS 2003. Table 3-1 presents the twenty-two

Table 3-1. *Mathematics Achievement and Student Age for Countries in TIMSS 1995 and 2003, Eighth Grade*

Country	Math 1995	Math 2003	Math change	Age 1995	Age 2003	Age change
Australia	507	505	−2	14.04	13.88	−0.16
Belgium (Flemish)	550	537	−13	14.14	14.12	−0.02
Cyprus	468	459	−8	13.74	13.77	0.03
England	498	498	1	14.05	14.29	0.25
Hong Kong SAR	569	586	17	14.18	14.39	0.21
Hungary	527	529	3	14.28	14.51	0.23
Iran	418	411	−7	14.63	14.44	−0.20
Japan	581	570	−11	14.38	14.40	0.01
Korea	581	589	8	14.20	14.60	0.40
Latvia	488	508	20	14.27	15.05	0.78
Lithuania	472	502	30	14.26	14.94	0.68
Netherlands	529	536	7	14.35	14.26	−0.09
New Zealand	501	494	−7	14.00	14.05	0.05
Norway	498	461	−37	13.89	13.80	−0.08
Romania	474	475	2	14.58	14.97	0.39
Russian Federation	524	508	−16	14.03	14.19	0.16
Scotland	493	498	4	13.70	13.68	−0.02
Singapore	609	605	−3	14.55	14.33	−0.22
Slovak Republic	534	508	−26	14.26	14.32	0.05
Slovenia	494	493	−2	13.82	13.90	0.08
Sweden	540	499	−41	14.93	14.89	−0.04
United States	492	504	12	14.23	14.23	0.01

Note: N = 22.

included countries.[23] The table also presents the mean country-level mathematics achievement scores and student age, and the changes in achievement and age from the 1995 TIMSS to the 2003 study. The achievement scores were computed from a mean of the five plausible values assigned to each student, and the results presented in table 3-1 agree perfectly with those presented in the TIMSS report.

The difference in the mean age of the participating students in 1995 and 2003 is quite substantial for Latvia and Lithuania and, to a somewhat smaller extent, for Korea and Romania. For Latvia and Lithuania the difference amounted to about nine months. Differences of this magnitude must have been caused by some systematic factor, even though it is not clear why these age differences appeared. For most of the countries, however, the differences in mean age of the students are quite small and are likely to have been caused by random factors having to do with when the assessments were conducted. The countries were allowed to conduct the field work any time from late April

Table 3-2. *Correlations between Achievement and Age, Eighth Grade*

Item	Math 1995	Math 2003	Math change	Age 1995	Age 2003	Age change
Math 1995	1.00					
Math 2003	0.93	1.00				
Math change	−0.12	0.26	1.00			
Age 1995	0.19	0.14	−0.12	1.00		
Age 2003	0.05	0.16	0.29	0.75	1.00	
Age change	−0.14	0.08	0.58	−0.01	0.65	1.00

Note: N = 22.

to early June, and it is likely that the decisions involved in when to conduct the survey caused the observed variation in mean student age.

Table 3-2 presents correlations between the mathematics achievement variables and the age variables at the country level. Mathematics achievement in TIMSS 1995 correlated .93 with math achievement in TIMSS 2003; the correlation of .75 for mean student age between the two surveys was also substantial. Within each survey there was no correlation between mean age and mean achievement. For TIMSS 1995 achievement correlated .19 with age, and for TIMSS 2003 the corresponding correlation was .16. Even though these correlations are marginally positive they are not significant. The correlation of .58 ($p < .005$) between the math change variable and the age change variable was highly significant, however.

This simple analysis thus yields the somewhat paradoxical result that within each survey there was no association between student age and student achievement at the country level, but there was a strong correlation between change in age and change in mathematics performance between 1995 and 2003. One explanation for this pattern is that the correlation between age and achievement is influenced by cultural and economic factors and by such matters as country differences in school starting age. These factors, which are omitted variables in the analysis of cross-sectional data, may conceal a true correlation between student age and achievement. However, the correlation between change in achievement and change in age within countries keeps these factors associated with countries constant, thereby allowing the correlation between age and achievement to appear.

A regression analysis of the math change variable on the age change variable gives an unstandardized regression coefficient of 38, which implies that an age change of one year is associated with an increase in achievement of 38 points. This estimate reflects the combined effect of becoming one chronological year older and going to school one more year, so it agrees reasonably well with the

difference in achievement of 30 points from grade seven to grade eight found
in TIMSS 1995. There is, however, a subtle difference between the meanings
of these estimates. The difference in level of achievement between grades seven
and eight results from the combined effect of a twelve-month chronological
year and a school year that is typically around nine months. In contrast, the
unstandardized regression coefficient is based on the observed age variation
within a school year, which, when expressed in terms of the expected effect of
a one-year difference, captures the combined effect of a twelve-month chronological year and twelve months of schooling. Assuming that the effect of
another month in school is twice as large as just becoming one month older,
the regression estimate of 38 points may be rescaled into an estimate that is
comparable with the estimate for comparing adjacent grades. The adjusted estimate is 32 points, making it very close to the estimate of 30 points for the comparison of adjacent grades.

Figure 3-1 presents a plot of the math change variable against the age change
variable. From the figure it may be seen that for Lithuania and Latvia the mean
age of the students was eight to nine months higher in TIMSS 2003 than in
TIMSS 1995, and for Lithuania achievement was 30 points higher in 2003,
while for Latvia it was 20 points higher. For most other countries the mean age
difference only amounted to a few months, and in most cases the achievement
change was smaller than in Lithuania and Latvia. However, if the results for
Lithuania and Latvia are excluded from the regression, the estimated coefficient
is 28, which is still a substantial relation.

Results for Grade Four

The TIMSS 1995 and 2003 studies also included samples of fourth-grade students, and fifteen countries participated in both studies. Are the results presented above for grade eight replicated with the data for the fourth-grade
students? Table 3-3 lists the countries included, along with their mathematics
achievement score and the mean age of the students in 1995 and 2003.

Most of the countries improved their level of mathematics achievement
between 1995 and 2003; the mean change was 10 points. The amount of change
varied considerably, however, with Latvia improving the most (47 points) and
Norway dropping the most (−25 points). The mean age was somewhat higher
in 2003 than in 1995, the difference amounting to 0.07 years. Here too, however, countries varied. In Latvia the average age of the students was 0.60 years
higher in TIMSS 2003 than in TIMSS 1995, while in Iran students were
0.10 years younger in 2003 than in 1995. With the exception of Latvia, these
age differences were not larger than might be expected considering the two-to-
three month time frame in which countries may conduct the assessments.
Table 3-4 presents the correlations among the variables.

Figure 3-1. *Change in Mathematics Achievement as a Function of Change in Mean Student Age between TIMSS 1995 and TIMSS 2003, Eighth Grade*

Math score change

Note: N = 22.

The pattern of correlations for grade four was highly similar to the pattern observed for grade eight. The correlation between age and achievement was low within each of the two assessments (.21 and .33 for 1995 and 2003, respectively), but the correlation between change in mathematics achievement and change in age was substantial and highly significant ($r = .63$, $p < .013$). While the correlation between age change and change in achievement was highly

Table 3-3. *Mathematics Achievement and Student Age for Countries in TIMSS 1995 and 2003, Fourth Grade*

Country	Math 1995	Math 2003	Math change	Age 1995	Age 2003	Age change
Australia	495	499	4	9.86	9.89	0.03
Cyprus	475	510	35	9.84	9.90	0.06
England	484	531	47	10.04	10.27	0.23
Hong Kong	557	575	18	10.14	10.24	0.10
Hungary	521	529	7	10.41	10.55	0.14
Iran	387	389	2	10.50	10.40	−0.10
Japan	567	565	−3	10.39	10.41	0.02
Latvia	499	536	37	10.46	11.05	0.59
Netherlands	549	540	−9	10.26	10.23	−0.03
New Zealand	469	493	24	9.98	10.03	0.05
Norway	476	451	−25	9.87	9.81	−0.06
Scotland	493	490	−3	9.71	9.70	−0.01
Singapore	590	594	4	10.31	10.33	0.02
Slovenia	462	479	17	9.87	9.78	−0.09
United States	518	518	0	10.19	10.24	0.06

Note: N = 15.

similar for both grades, the estimate of the regression coefficient was higher in grade four ($b = 71$). However, this estimate too expresses the combined effect of a twelve-month chronological year and a twelve-month school year, so to be comparable to the estimate obtained from the comparison of adjacent grades (57 points), it should be adjusted in the same manner as was done for grade eight. The adjusted estimate is 59, so for grade four too the estimates agree excellently.

Figure 3-2 presents the plot of change in achievement against change in age for the fourth-grade sample. The plot clearly indicates the high level of corre-

Table 3-4. *Correlations between Achievement and Age, Fourth Grade*

Item	Math 1995	Math 2003	Math change	Age 1995	Age 2003	Age change
Math 1995	1.00					
Math 2003	0.93	1.00				
Math change	−0.16	0.22	1.00			
Age 1995	0.21	0.21	−0.01	1.00		
Age 2003	0.22	0.33	0.29	0.89	1.00	
Age change	0.14	0.38	0.63	0.34	0.72	1.00

Note: N = 15.

Figure 3-2. *Change in Mathematics Achievement as a Function of Change in Mean Student Age between TIMSS 1995 and 2003, Fourth Grade*

Math score change

Note: N = 15.

lation between the two change variables. It also indicates that Latvia is somewhat of an outlier, with a very high improvement in achievement and a high change in mean age. It may be asked, therefore, if the high correlation between the two variables is due to the results for this particular country. However, when the correlations are recomputed with Latvia excluded, the same result is obtained ($r = .63, p < .015$).

Summary

In theory, the mean student age should be the same for countries participating in TIMSS 1995 and TIMSS 2003, because samples are drawn from the same grade and the testing should be conducted at the same time of the year. There is, however, some variation in testing time, and even though this variation may be random, the existing variation is related to achievement. And if the variation in student age between the different cohorts is random, it is not correlated with other variables, which makes inferences about the effect of age on achievement unaffected by bias caused by omitted variables.

Analyses of the relation between student age and achievement within each of the two cross-sectional data sets did not show any correlation between age and achievement. This indicates that these analyses are influenced by omitted variables that disturb the positive relation between age and achievement. An important determinant of the age variability at the time of testing is the school start age, which varies over groups of countries. (The Nordic countries, for example, have a high school start age.) It is thus reasonable to assume that this variable is associated with a large set of cultural and educational factors, which may bias the relations between age and achievement.

The analysis of the relation between change in achievement and change in mean student age between 1995 and 2003 gives an estimate of the effect of student age that agrees with results from other approaches. This indicates that a longitudinal analysis aggregated to the country level is a simple method of controlling for the influence of omitted variables, and it is reasonable to assume that this is because the differences in mean student age between the two assessments are not systematically related to any other variable.

Example 2: Effects of Class Size on Student Achievement

Let me now turn to a quite different example. Most parents, teachers, and students expect learning to be more efficient in a smaller class than in a larger class. Furthermore, class size is one of the most important variables for determining the level of resources needed for an educational system. Many studies have been conducted on the effects of class size on achievement, but the results have tended to be inconsistent, and at least up to the early 1990s the general conclusion was that neither resources in general, nor class size in particular, have any relation to achievement.

During the 1990s, however, several influential studies demonstrated that class size does indeed matter. One of these studies, known as the STAR experiment (Student/Teacher Achievement Ratio), started in 1985. This large-scale study had three treatment groups: small classes with 13–17 students; regular classes with 22–26 students; and regular classes with an assistant teacher. To be included in the study, a school had to have at least one class of each type. Some 80 schools participated; altogether there were more than 100 classes of each type. During the first year of the study, about 6,000 students were included; by the end of the fourth year almost 12,000 students were involved, because of addition of new students. Within schools both students and teachers were randomly assigned to the three treatments. Most of the students entered the study either in kindergarten or first grade, while a few entered in second or third grade. In the first phase of the study, the students were followed through third grade, with measures of achievement being made at the end of each grade. The great strength of this design is, of course, that the randomization of both students and teachers over the three treatment conditions implied that there were no problems of causal inference due to selection bias or omitted variables.

At the end of grade three, the results showed a striking advantage for the students who had been assigned to the small classes, while there was no clear difference between the results achieved in regular classes with one teacher, and regular classes that also had an assistant teacher.[24] The results were particularly strong for reading, with an effect size of around .25, and it was found that the small-class advantage was larger for students who came from socioeconomically and ethnically disadvantaged groups.[25] The results showed the effect of class type to be strongest for first grade, while the difference kept more or less constant over second and third grades.

The results from the STAR experiment have been replicated in other studies.[26] Robinson reported a large meta-analysis of more than 100 studies of class size.[27] The results indicated that the effect of class size interacts with the age of the students. Robinson found that for kindergarten and the first three grades, small classes have a positive effect; for grades four through eight, they have a weak positive effect. Studies conducted on students in grades nine through twelve provided no support for the hypothesis that class size has an effect on achievement.

Other studies have relied on natural variation in class size, using sophisticated statistical techniques to try to sort out causal effects of class size from selection effects. Angrist and Lavy conducted a study in Israel, where a strict rule on the maximum number of students in a class implies that one large class is split into two smaller classes.[28] Because the variation in class size caused by the splitting rule is essentially unrelated to factors such as the socioeconomic status of the area that the school is recruiting from, the variation can be used to estimate effects of class size. Using so-called instrumental variable estimation on data from some 2,000 fourth- and fifth-grade classes, class size was found to have a significant effect on achievement in reading and mathematics in fifth grade, and in reading in fourth grade, with smaller classes producing the better results. The effect sizes were somewhat lower than those found in the STAR experiment (for example, .18 for grade five, assuming the small classes had eight fewer students than the large classes) but were still judged large enough to be practically important. As in the STAR experiment, Angrist and Lavy also found an interaction between socioeconomic background and class size, with the benefits of small classes being larger in schools with a large proportion of students from a disadvantaged background.

Hoxby took advantage of the fact that natural variation in population size causes random variation in class size that is not associated with any other variation, except perhaps achievement.[29] Like Angrist and Lavy, she too investigated the abrupt changes in class size caused by rules about maximum class size. Using data from 649 elementary schools covering a period of twelve years, Hoxby found no significant effect of class size on achievement, even though she demonstrated that she was able to detect class-size effects as small as those found in the experimental research. Hoxby suggests that the differences between the experimental studies and her studies of natural variation might have had to do with the teachers in the experimental studies, who may have preferred small classes and thus tried harder to produce improved student achievement in them.

Several studies have used the IEA data, and TIMSS in particular, to investigate effects of class size and other resource factors. Hanushek and Luque used data from the TIMSS 1995 study to investigate effects of resources on educational achievement, focusing on possible differences in effects between different countries.[30] They analyzed data for 9-year-olds (population A) and 13-year-olds (population B), modeling relations at the school level within each country between resource and background factors on the one hand and achievement on the other hand.

One general finding that emerged from the analyses was that effects of resources seemed to be stronger in other countries than in the United States. However, there was little consistency over countries and age groups in the pattern of results. For class size, for example, fourteen out of seventeen estimated relations with achievement for the age 9 samples had the expected negative sign, while twenty-three out of thirty-three estimated relations for the age 13 samples had a positive sign. This pattern suggests that smaller classes are beneficial for the achievement of young students, but not for older students—a pattern that agrees quite well with the findings in the research literature that any positive effect of smaller class size is restricted to the first years of schooling.

Hanushek and Luque were suspicious, however, about the finding that in the majority of countries there was a positive effect of being in a larger class for 13-year-olds. They hypothesized that this result might have been an effect of selection bias caused by a tendency to place weaker students in smaller classes, and they tested this hypothesis in two different ways. In one approach they performed a separate analysis of schools in rural areas, arguing that such schools typically have only one classroom, leaving no room for any mechanism of selection bias to operate. However, this analysis produced the same result found for the complete set of schools. In the other approach the researchers asked the principal whether the class size in the sampled classroom was below average for the grade in the school. The results showed that in five of thirty-two countries, the 13-year-olds in the smaller classrooms had lower achievement than the grade average for the school, which supports the hypothesis that weaker students tend to be placed in smaller classes. According to Hanushek and Luque, however, the estimated class-size effects did not change even when such within-school compensatory placements were taken into account.

These results thus indicate that for the older age group, class size does not seem to be related to student achievement, while for the younger age group, being in a small class may improve achievement.

Wößmann and West also analyzed the TIMSS 1995 population B data to determine effects of class size.[31] To come to grips with the problem of selection bias within and between schools, they used a sophisticated instrumental variable estimation approach, which relied upon differences in class size between adjacent grades, and differences between mean class size of the school and the actual class size. The analysis also took into account country fixed effects. Adequate data to perform the estimation was available for eleven countries. The estimation results showed statistically significant negative effects of class size on achievement in a few cases (France and Iceland in mathematics, and Greece and Spain in science), while in the vast majority of cases class size had no effect. Thus, this study too indicates that for students in grades seven and eight, class size is not an important determinant of achievement.

Summarizing the results from these studies, it appears that smaller classes may have some positive effect on achievement during the first few years of schooling, but little or no effect after that. There are, however, many studies in which the results run against this generalization, so further research is necessary.

Results for Grade Eight

Results are reported below from analyses in which change in class size has been related to change in mathematics achievement between 1995 and 2003 for grades four and eight. These analyses rely on the same set of countries in TIMSS 1995 and 2003 that were analyzed in the previous example. However, unless class sizes have changed in the participating countries, there is no reason to investigate correlates of this variable. Class size at the country level has been estimated from information elicited in the teacher questionnaire about the size of mathematics classes. For the 1995 data there were separate variables for the number of boys and girls in the class, while for the 2003 data there was a variable providing the total number of students in each class. This difference,

Table 3-5. *Correlations between Achievement and Class Size, Eighth Grade*

Item	Math 1995	Math 2003	Math change	Class size 1995	Class size 2003	Class size change
Math 1995	1.00					
Math 2003	0.93	1.00				
Math change	−0.12	0.26	1.00			
Class size 1995	0.42	0.47	0.16	1.00		
Class size 2003	0.52	0.60	0.25	0.87	1.00	
Class size change	−0.03	0.00	0.08	−0.64	−0.17	1.00

Note: N = 22 .

which applied only to grade eight, may have introduced some extra variability in the estimates of change of class size, given that the distributions of class size for boys and girls displayed some peculiarities.

In 1995 the international mean class size for grade eight was 27.3; in 2003 the international mean was 26.1, so class sizes were growing smaller. The standard deviation of the class size change was 3.7, indicating considerable variability in the amount of change over the countries. The correlations among the variables are presented in table 3-5.

The table shows no correlation ($r = .08$, $p < .71$) between change in class size and change in achievement. This result thus agrees with several other studies showing no causal effect of class size on achievement for older students. The cross-sectional data, however, show a significant positive relationship between class size and achievement in mathematics both in 1995 ($r = .42$, $p < .050$) and in 2003 ($r = .60$, $p < .003$). These results indicate that larger classrooms produce a higher level of achievement than smaller classrooms. However, the class-size variable may be correlated with a large number of other variables that may be instrumental in causing a higher level of achievement, so the cross-sectional analysis may be influenced by omitted variable problems.

Figure 3-3 presents a plot for the twenty-two countries of mean mathematics achievement and mean class size. This plot shows that the positive correlation is caused by the high level of performance and the large class size of a group of four countries, namely, Hong Kong, Japan, Korea, and Singapore. These countries are all East Asian countries, and no other country in the study belongs to this group. In addition to larger class sizes, these Asian countries display many other educational differences from other countries in the world, including a strong cultural emphasis on education and math, a high level of teaching quality, carefully planned lessons, and using larger classes to advantage.[32] These differences suggest that country differences in a complex set of cultural and educational factors may account for the positive relation between

Figure 3-3. *Plot of Mean Mathematics Achievement as a Function of Mean Country Class Size in TIMSS 2003, Eighth Grade*

Math score 2003

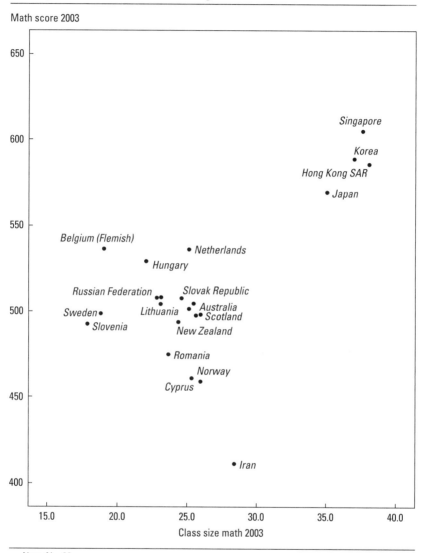

Note: N = 22.

country-level class size and math achievement rather than country differences in class size per se.

The absence of a zero-order correlation between change in class size and change in mathematics performance makes it unlikely that bringing further variables into the analysis will yield any other results. But the substantial correlation between change in age and change in mathematics achievement may make it worthwhile to include both age change and class-size change as independent variables in a regression analysis. According to this analysis, the partial regression coefficient for class-size change was almost exactly 0, and the regression coefficient for age was 38, which was also the estimate that was obtained when the variable was entered alone into the model. As has been the case in many other studies, this analysis thus provides no ground for claiming that class size has an effect on mathematics achievement for eighth-grade students.

Results for Grade Four

Class size for fourth-grade students also declined for the fifteen participating countries, with the international mean class size dropping slightly, from 26.3 in the TIMSS 1995 study to 25.7 in 2003. The standard deviation of the class size change was 2.6, which indicates variability over the countries with respect to the amount of change. The correlations between the variables are presented in table 3-6.

Cross-sectional data for fourth graders also showed positive correlations between mathematics achievement and class size for 1995 and 2003 (.34 and .54, respectively). The class-size change variable correlated −.22 with change in mathematics achievement, which was not significant ($p < .44$). Adding the age change variable, however, caused both the age change variable and the class-size change variable to be highly significant ($t = 4.91$ and $t = −3.32$, respectively); these two variables account for 68 percent of the variance in the mathematics change variable.

The regression coefficient was −4.44 for the regression of change in mathematics achievement on class-size change, which implies that a class-size reduction of seven students, as in the STAR experiment, would yield an improvement of 31 points. This improvement roughly translates into an effect size of .31. According to Finn and Achilles, the effect size of being in a small class at the end of third grade was .26 for reading and .23 for mathematics.[33] These effect sizes were quite different for different groups of students, however. For mathematics, the effect size was .16 for white children but .30 for minority children. Thus, even though the effect size in this analysis seems somewhat higher than was found in the STAR experiment, the results do not seem to be contradictory.

Summary

These analyses suggest that the TIMSS 1995 and 2003 eighth-grade data show no effect of class size on academic achievement, because change in class size at the country level was not related to change in achievement over time. It also could be concluded that the strong cross-sectional correlation between achievement and class size at the country level on both occasions was due to the fact that

Table 3-6. *Correlations between Achievement and Class Size, Fourth Grade*

Item	Math 1995	Math 2003	Math change	Class size 1995	Class size 2003	Class size change
Math 1995	1.00					
Math 2003	0.93	1.00				
Math change	−0.16	0.22	1.00			
Class size 1995	0.34	0.39	0.13	1.00		
Class size 2003	0.53	0.54	0.05	0.90	1.00	
Class size change	0.25	0.16	−0.22	−0.52	−0.10	1.00

Note: N = 15.

both these variables had high values for a group of four East Asian countries. It must be emphasized, of course, that the negative results for grade eight should not be interpreted as proof that there is no effect of class size on educational achievement. The correctness of the null hypothesis cannot be proved, and there may be alternative explanations for the lack of relationship between class size change and achievement change. For example, a slight change in the definition of the class-size variable between the 1995 data and the 2003 data was noted.

For the fourth-grade data, the trend analysis between 1995 and 2003 showed a significant effect of class-size change on change in mathematics achievement, with the expected negative relationship between class size and achievement. The size of the estimated effect, though somewhat higher, agrees reasonably well with findings in previous research.

The general pattern of findings emerging from these two sets of analyses for populations A and B agrees quite well with the conclusions from the cumulated research on the effects of class size, namely, that class size does not affect achievement for students in grade six and higher, but that it is of importance for primary school students. This agreement should not be interpreted as proof that the analytical procedure focusing on change over time at the country level provides unbiased estimates of causal relations, but the results are encouraging and do support further work along these lines.

Discussion and Conclusions

The main aim of this chapter was to identify possible threats to the correctness of causal inference from the international studies of educational achievement and to suggest approaches for avoiding those threats. Selection bias (or reverse causality or endogeneity problems) and omitted variables are two major sources of problems. These problems seem almost impossible to deal with through analyses at low levels of aggregation of cross-sectional data. It was therefore suggested that data be analyzed at higher levels of aggregation focusing on change over time for the same units of observation. Given that the latest generation of international studies has been designed to yield information about trends in the development of knowledge and skills within countries, aggregation could be made at the country level, and an analytical approach taking advantage of the longitudinal design could be adopted. There should be no mechanisms generating selection bias at the country level, and the fact that change over fixed countries is analyzed turns many of those factors that vary over countries into constants so that they cannot correlate with the independent variables under study.

To investigate the tenability of these ideas, two examples were studied using data from grades four and eight in TIMSS 1995 and 2003, namely, the effects of age and class size on achievement. Both these examples demonstrated that the analyses based on cross-sectional data yielded biased results, while the results from the longitudinal country-level analyses were reasonable and compatible with results obtained in studies using other methodological approaches. Country-level trend analyses may thus be a useful addition to the set of tools available for secondary analysis of the data from the comparative international studies.

It must be emphasized, though, that this analytical approach is not a panacea, and that many problems cannot be studied with this approach. It also is easy to envision threats to the validity of causal inferences from country-level longitudinal analyses.

Problems that focus upon explanatory variables that do not change over time obviously cannot be studied with the longitudinal approach proposed here. For example, the institutional factors investigated by Wößmann, such as the use of centralized examinations, are not likely to change over shorter periods of time, or at least not to such an extent that any effects can be determined.[34] Other approaches are needed for investigating the effects of institutional factors at the country level (although the microeconometric modeling approach applied by Wößmann does not seem to avoid the omitted variable problem).

There is, of course, no guarantee that change measures of explanatory variables are uncorrelated with other independent variables, and that they therefore are not afflicted by the omitted variable problem. The age change variable investigated in the first example is a more or less random variable, at least for grade four. This makes it optimal as an explanatory variable, because if it is random, it is not correlated with other independent variables, even though it still exerts an influence on change in achievement. But for most other independent variables of any educational interest, we cannot assume that change is uncorrelated with other variables. Suppose, for example, that a set of countries that have achieved a poor result in a comparative study take different measures to improve achievement, such as lowering class size and improving teacher competence. Sorting out the effects of the various resource factors in such a case is likely to be difficult. To add to the analytical complexity, this scenario is likely to produce some regression toward the mean of the country results when the assessment is repeated. As always, great care must thus be taken in analyzing and interpreting the results.

In this paper the simplest possible analytical techniques have been applied to keep the focus on the basic methodological issues. Correlation and regression analysis of difference scores can be applied in a meaningful manner, however, only when there are two waves of measurement. For TIMSS, which is now conducted every four years, three waves have already been completed, and a fourth wave of data collection is being conducted in 2007. For the Progress in International Reading Literacy Study (PIRLS), the second wave of measurement was completed in 2006, and the third will be completed in 2011. To take full advantage of such multiwave data, other techniques should be employed. One technique that would seem to be a natural choice when a systematic trend over a longer period of time is expected is growth modeling at the country level.

Analysis of trend data can be extended in many other interesting ways. The total sample of students, for example, could be broken into results for different subgroups, such as gender, language spoken at home, and socioeconomic background. It also might be interesting to investigate variability of scores as an outcome when questions of equity are investigated.

While the paper focuses primarily on methodological issues, the results concerning class size are of substantive interest, even though that will not be further discussed here. The results concerning effects of age differences at time of testing for the different waves of measurement are of limited theoretical interest, but they do seem to be of practical interest in the design of the implementation of the assessments. It has been demonstrated here that these age differences accounted for 30–40 percent of the variance in the change in achievement between 1995 and 2003. This is a substantial share of the total variance, and there may be reason to try to reduce it, perhaps by narrowing the time frame in which the assessments can be conducted. Alternatively, the age differences could be taken into account through statistical adjustment.

As noted at the beginning, IEA was founded with the intention of using the world as an educational laboratory. While the founding fathers might have been overly optimistic about the possibilities for overcoming the technical and methodological problems they would encounter, many problems, such as those associated with sampling and measurement, have successfully been mastered. The laboratory is still somewhat messy, however, and in particular the problem of how to sort out the multitude of factors influencing achievement remains to be solved. Systematic analysis of trends of achievement in different countries may provide at least a partial solution to this problem. Further work will be needed to determine to what extent this is true.

Notes

1. Torsten Husén, ed., *International Study of Achievement in Mathematics: A Comparison of Twelve Countries,* 2 vols. (New York: Wiley, 1967).
2. Albert E. Beaton and others, *Mathematics Achievement in the Middle School Years: IEA's Third International Mathematics and Science Study (TIMSS)* (Boston College, 1996); Ina V. Mullis and others, *Mathematics Achievement in the Primary School Years: IEA's Third International Mathematics and Science Study (TIMSS)* (Boston College, 1997).
3. Ina V. Mullis and others, *TIMSS 2003 International Mathematics Report. Findings from IEA's Trends in International Mathematics and Science Study at the Fourth and Eighth Grades* (Boston College, 2004).
4. António Novóa and Tali Yariv-Mashal, "Comparative Research in Education: A Mode of Governance or a Historical Journey?" *Comparative Education* 39, no. 4 (November 2003): 423–38.
5. Lyle V. Jones and Ingram Olkin, eds., *The Nation's Report Card. Evolution and Perspectives.* (Bloomington, Ind.: Phi Delta Kappan, 2004).
6. Organization for Economic Cooperation and Development, *Knowledge and Skills for Life: First Results from PISA 2000* (Paris: 2001).
7. Erik Allardt, "Challenges for Comparative Social Research," *Acta Sociologica* 33, no. 3 (1990): 183–93.
8. Trevor Williams and others, "Achievement and Affect in OECD Countries," *Oxford Review of Education* 31, no 4 (December, 2005): 517–45.
9. Ina V. Mullis, Jay R. Campbell, and A. E. Farstrup, *NAEP 1992 Reading Report Card for the Nation and the States: Data from the National and Trial State Assessments,* NCES Report 23-ST06 (U.S. Department of Education, September 1993).
10. Edward P. Lazear, "Educational Production," *Quarterly Journal of Economics* 116, no. 3 (August 2001): 777–803.
11. Ludger Wößmann, "Schooling Resources, Educational Institutions, and Student Performance: the International Evidence," *Oxford Bulletin of Economics and Statistics* 65, no. 2 (2003): 117–71.
12. Ibid., p. 124.
13. Ibid., p. 120.
14. Ibid., p. 121.
15. Sorel Cahan and Nora Cohen, "Age versus Schooling Effects on Intelligence Development," *Child Development* 60, no. 5 (1989): 1239–49.
16. Jan-Eric Gustafsson, "What Do We Know about Effects of School Resources on Educational Results?" *Swedish Economic Policy Review* 10, no. 2 (2003): 77–110
17. Lazear, "Educational Production."
18. Ludger Wößmann and Martin West, "Class-Size Effects in School Systems around the World: Evidence from Between-Grade Variation in TIMSS," *European Economic Review* 50, no. 3 (2006): 695–736.
19. Anthony S. Bryk and Stephen W. Raudenbush, *Hierarchical Linear Models: Applications and Data Analysis Methods.* (Newbury Park: Sage Publications, 1992); Linda Muthén and Bengt O. Muthén, *Mplus User's Guide,* 4th ed. (Los Angeles, Cal.: Muthén & Muthén, 2006).
20. Erik A. Hanushek and Ludger Wößmann, "Does Educational Tracking Affect Performance and Inequality? Differences-in-Differences Evidence across Countries." CESifo Working Paper 1415 (Munich: Institute for Economic Research, February 2005).
21. Beaton and others, *Mathematics Achievement in the Middle School Years,* p. 28.
22. Cahan and Cohen, *Effects of Age versus Schooling on Intelligence,* p. 1246.
23. This list includes one country fewer than was included in the analysis of trends between 1995 and 2003 presented in the TIMSS report, namely, Bulgaria. The reason is that the data collected in 1995 for Bulgaria lack most of the background questionnaires, which makes this country of limited interest for the current analytical purposes.

24. Jeremy D. Finn and Charles M. Achilles, "Answers and Questions about Class Size: A Statewide Experiment," *American Educational Research Journal* 28, no. 3 (Autumn 1990): 557–77; Jeremy D. Finn and Charles M. Achilles, "Tennessee's Class Size Study: Findings, Implications and Misconceptions," *Educational Evaluation and Policy Analysis* 21, no. 2 (1999): 97–110.

25. Alan B. Krueger, "Experimental Estimates of Educational Production Functions," *Quarterly Journal of Economics* 114, no. 2 (1999): 497–532.

26. For a review, see Gustafsson, "What Do We Know about Effects of School Resources."

27. Glen E. Robinson, "Synthesis of Research on the Effects of Class Size," *Educational Leadership* 47, no. 7 (1990): 80–90.

28. Joshua D. Angrist and Victor Lavy, "Using Maimonides' Rule to Estimate the Effect of Class Size on Scholastic Achievement," *Quarterly Journal of Economics* 114, no. 2 (1999): 533–75.

29. Caroline Hoxby, "The Effects of Class Size on Student Achievement: New Evidence from Population Variation," *Quarterly Journal of Economics* 115, no. 4 (2000): 1239–85.

30. Eric A. Hanushek and Javier A. Luque, "Efficiency and Equity in Schools around the World," *Economics of Education Review* 22, no. 5 (2003): 481–502.

31. Wößmann and West, "Class Size Effects in School Systems around the World."

32. John Biggs, "Learning from the Confucian Heritage: So Size Doesn't Matter?" *International Journal of Educational Research* 29, no. 8 (1998): 723–38.

33. Finn and Achilles, "Tennessee's Class Size Study."

34. Wößmann, "Schooling Resources, Educational Institutions, and Student Performance."

4

Lack of Focus in the Mathematics Curriculum: Symptom or Cause?

WILLIAM H. SCHMIDT AND RICHARD T. HOUANG

Two reports, released in conjunction with the original TIMSS (Third International Mathematics and Science Study) in 1997, analyzed the curriculum component of the study and offered one significant explanation for the relatively poor performance of U.S. students in mathematics.[1] The data indicated that the U.S. curriculum typically covered more topics at each grade level than did that of any other country participating in TIMSS—leading to the description of the U.S. math curriculum as one that was "a mile wide and [an] inch deep." This characterization was true of state curriculum standards, the National Council of Teachers of Mathematics (NCTM) standards, district standards, textbooks, and even the curriculum as implemented by the teachers. There was little focus in the U.S. curriculum, given the large number of topics that were intended to be, and actually were, covered. In this paper, we return to the issue of curriculum focus and ask whether such a concept is useful, not just as a hypothesis about the United States but, more generally, in understanding cross-national differences in achievement.

Since 1997, when these original observations were made, greater insight has been gained into the cross-national differences in curriculum and their relationship to achievement.[2] One such insight has been labeled coherence. It is applicable to different aspects of the curriculum, but here we examine coherence in curriculum content standards:

> Content standards, taken together, are coherent if they are articulated over time as a sequence of topics and performances consistent with the logical and, if appropriate, hierarchical nature of the disciplinary content from which the subject matter derives. This is not to suggest that there is only one coherent sequence, but rather that any such coherent sequence reflect[s] the inherent structure of the discipline. This implies that for a set of content standards "to be coherent" they must evolve from particulars (e.g., simple mathematics facts and routine computational procedures associated with whole numbers and fractions) to deeper structures. It is these deeper structures by which the particulars are connected (such as an understanding of the rational number system and its properties). This evolution should occur both over time within a particular grade level and as the student progresses across grades.[3]

If content standards reflect the structure of a discipline, then the "depth" of those standards should *increase* as students move across the grades. Failure to increase in depth, sophistication, and complexity across the grades would indicate a lack of coherence. Extensive repetition of virtually the same standards across grade levels is found in the United States. This repetition runs counter to the idea of "coherent" development, is unwarranted, and contributes to a lack of focus. Such repetition can be replaced with standards that form a trajectory by linking coverage of the topics over grades and by reducing the repetition over the same grade levels. Such an approach would represent a "continuing penetration of the discipline moving to a deeper structure that makes things 'simpler.'"[4] That is, coherent development is, in the long run, fundamentally simpler than virtual repetition without development in depth and sophistication and without attaching to fundamental, unifying ideas of the discipline.

Once one begins to consider the concept of coherence, one could conclude that focus is merely a symptom or indicator and that its absence suggests that coherence is lacking in the curriculum. In this interpretation, the lack of coherence is the real difficulty. The lack of focus, the large number of topics in a curriculum, is a symptom, like a high temperature, of a problem and not the problem itself.

Some have argued that focus is not important at all. For example, Baker and LeTendre contend that the move toward globalization includes globalization of the curriculum. They suggest that most nations now intend the "same knowledge base in mathematics and science, significantly reducing cross-national variation in the intended curriculum."[5] Using the concepts of breadth and depth, which parallel our concept of focus, they find only small differences in both dimensions (as they define them) across the TIMSS countries and no significant relationship between breadth and depth of curricula and cross-national differences in achievement.

There is a problematic clash of views about focus and its significance. Is focus only a symptom of what may be problematic in characterizing curriculum among nations? Is it a root cause related to achievement differences? Is it neither—that is, is focus not important at all either as a symptom in characterizing country content standards or as a factor in the relationship of content standards to cross-national differences in achievement?

The Data

The 1995 TIMSS was the most complex and extensive cross-national comparative study of education that had ever been attempted.[6] As a part of this study, the curriculum component analyzed the national standards and textbooks of almost fifty countries, using document analysis coding procedures. The coding was based on a cross-nationally developed and tested consensual framework that represented the topics in mathematics covered in grades one through eight in the TIMSS countries.[7]

The content standards in mathematics of the top-achieving TIMSS countries were subsequently used to develop a model of coherence. A methodology was developed to derive a set of international standards, which were then analyzed by research mathematicians and found to be consistent with the inherent logic of mathematics. For the purpose of this study, those international benchmarks become our empirical, logical definition of coherence in mathematics.

The six top-achieving countries were used to define coherence in mathematics over grades one to eight—Belgium (Flemish), the Czech Republic, Hong Kong, Japan, Korea, and Singapore. Schmidt, Wang, and McKnight describe the procedure as follows:

> The data used to develop the international benchmarks come from the curriculum component of TIMSS and derive from the procedure known as General Topic Trace Mapping (GTTM). . . . The respondents to the GTTM were education officials (typically curriculum officers in the national ministry) of each nation who, using their national content standards or an aggregate of regional standards, indicated for each grade level whether or not a content topic was included in their country's intended curriculum. The result was a map reflecting the grade-level coverage of each topic for each country. . . .
>
> Topic trace maps were available for each of the [top-achieving] countries. While none were identical they all bore strong similarities. The following procedures were followed to develop an international benchmark. First, the mean number of intended topics at each grade level was determined across the countries. Next, the topics were ordered at each grade level based on the percentage of [top-achieving] countries that included it in their curriculum. Those topics with the greatest percentage were chosen first, and only as many were chosen as were indicated by the mean number of intended topics at that grade level.[8]

The United States does not have a national curriculum, so for purposes of this paper, we drew a sample of twenty-one state standards to represent the United States in the cross-national analysis, using probabilities proportional to the size of the school population of the state. These were the content standards in place at the time of the TIMSS study. The state standards we sampled and the NCTM standards (which continue to have a strong influence on state standards) were subjected to a slightly different and more extensive content analysis coding procedure than that used for the development of the inter-national benchmarks. The TIMSS curriculum analysis component developed an elaborate document analysis methodology that was used to generate the U.S. data using the twenty-one state standards.

The Curriculum Structure of the Top-Achieving Countries

Table 4-1 portrays the set of topics for grades one through eight that represents only the common topics intended for coverage by two-thirds or more of the top-achieving countries. The data suggest a hierarchical sequencing moving from elementary to more advanced topics, resulting in a three-tier pattern of increasing mathematical complexity:

> The first tier, covered in grades 1–5, includes an emphasis primarily on arith-metic, including whole-number concepts and computation, common and decimal fractions, and estimation and rounding. The third tier, covered in grades 7 and 8, consists primarily of advanced number topics, algebra, includ-ing functions and slope, and geometry, including congruence and similarity and 3-dimensional geometry. Grades 5 and 6 appear to serve as an overlap-ping transition or middle tier marked by continuing attention to arithmetic topics (especially fractions and decimals), but with an introduction to the top-ics of percentages, negative numbers, integers, proportionality, co-ordinate geometry, and geometric transformations.[9]

The progression of the topics over grades appears to be based mostly on the inherent logic of the mathematics and is logically consistent with the nature of mathematics. It must be pointed out that this model of coherence—derived empirically and confirmed logically—does not represent the curricular map of any country, not even of the six from which its structure was developed. It is an estimate of what is typical among the pool of top-achieving countries.

Method

We wished to explore the issue of whether focus is itself an important and sep-arate indicator of the curriculum or whether it is essentially redundant with coherence (that is, a lack of focus is merely a reflection of a lack of coherence). We developed statistical indicators of both concepts. The GTTM data for

Table 4-1. *Mathematics Topics Intended at Each Grade by a Majority of TIMSS 1995 Top-Achieving Countries at Eighth Grade*

	Grades							
Topic	*1*	*2*	*3*	*4*	*5*	*6*	*7*	*8*
Whole number: meaning	●	●	●	●	●			
Whole number: operations	●	●	●	●	●			
Measurement units	●	●	●	●	●	●	●	
Common fractions			●	●	●	●		
Equations & formulas			●	●	●	●	●	●
Data representation & analysis			●	●	●	●		●
2-D geometry: basics			●	●	●	●	●	●
2-D geometry: polygons & circles				●	●	●	●	●
Measurement: perimeter, areas, & volume				●	●	●	●	●
Rounding & significant figures				●	●			
Estimating computations				●	●	●		
Whole numbers: properties of operations				●	●			
Estimating quantity & size				●	●			
Decimal fractions				●	●	●		
Relation of common & decimal fractions				●	●	●		
Properties of common & decimal fractions					●	●		
Percentages					●	●		
Proportionality concepts					●	●	●	●
Proportionality problems					●	●	●	●
2-D geometry: coordinate geometry					●	●	●	●
Geometry: transformations						●	●	●
Negative numbers, integers, & their properties						●	●	
Number theory							●	●
Exponents, roots, & radicals							●	●
Exponents & orders of magnitude							●	●
Measurement: estimation & errors							●	
Constructions using straightedge & compass							●	●
3-D geometry							●	●
Geometry: congruence & similarity								●
Rational numbers & their properties								●
Patterns, relations, & functions								●
Proportionality: slope & trigonometry								●

Source: TIMSS data, 1997.

● Intended by more than half of the top-achieving countries.

each country were mapped into the matrix defined by table 4-1. The rows and columns of table 4-1 were considered as fixed and remained constant. The resulting map for each country placed an indicator (represented graphically by a dot, as shown in table 4-1) into various cells defined by the 32 by 8 matrix (256 possible cells or opportunities to cover mathematics topics).

Those dots indicated that the topic defined by the row was covered in the content standards of that country in the grade defined by the column. In effect, this map indicated what topics were covered at which grades for each country.

The region of the matrix with dots in table 4-1 was taken as a model or ideal scenario defining coherence. For each country, the number of "hits" within the model region was considered an indicator of the degree of coherence. Here is one way to think of this process. The model region was highlighted, creating a "silhouette," as shown in table 4-1. This silhouette was then superimposed on the country maps. Then, the degree of overlap was used to estimate the coherence of that country's curriculum. A high value on this statistic indicates that the part of the curriculum map of the country dealing with the silhouetted region is very similar to the ideal scenario. As discussed previously, we are not arguing for only one model of coherence. Coherence could take another form. Low values on the proposed indicator should not necessarily be interpreted in the absolute as a lack of coherence but rather as a deviation from the empirically derived ideal scenario of coherence presented in table 4-1. It should be remembered that the table is a composite of the top-achieving countries and does not represent the curriculum of any individual country, including those countries that contributed to its definition. In this way, the silhouette serves a role similar to that played by a mean in other statistical analyses.

Since coherence is cumulative over the grades, its definition at specific grade levels demands that the silhouette in table 4-1—what we call the "upper triangle"—must be correspondingly partitioned. Through grade three, thirteen cells are silhouetted in our coherence model; through grade four, twenty-eight cells; through grade seven, eighty-one cells; and through grade eight, ninety-nine cells, as represented by the shaded area of the table.

The statistic used as an indicator of focus was a cumulative count of the total number of dots in the country map up to and including the grade in question. Large values indicated a lack of focus—typically the result of including a large number of topics in grades outside those needed to reproduce the empirical pattern of coherence.

For any country, the 256 cells in the matrix can be divided into three groups:

1. Cells that overlap with the upper triangle area—those that reflect an exact match with the ideal scenario of coherence; a count of these is used as the indicator of coherence.

2. Cells before the upper triangle—those topics that are introduced too soon compared with the ideal scenario.

3. Cells after the upper triangle—those topics that are introduced or continued after coverage in the ideal scenario is finished.

The sum of the three groups is used to define focus. It indicates the total number of cells or topic and grade combinations intended to be covered by the country up to and including that grade. It is a cumulative index and as a result the matrix must be partitioned appropriately for each separate grade.

A second statistic for estimating focus is based on a count of the number of intended topic and grade combinations before the upper triangle. A positive value for this indicator of intended coverage (number 2 above) reflects a lack of focus created by adding more topics to earlier grades. This particular lack of focus implies an early treatment of topics where the necessary prerequisite mathematics would likely be simultaneously covered or, what is even more problematic, not covered until a later grade. Examples of both kinds are found in the data. The latter, we argue, is the more serious.

The final estimator employed in the analyses is one that combines the concepts of coherence and focus in the same statistic. It is the ratio of the estimate of coherence and the estimate of focus. The total number of topic and grade combinations or cells that match the upper triangle area is divided by the total number of topic and grade combinations intended up to the particular grade in question. For an individual country, this indicates the proportion of intended opportunities to cover topics that are covered in the ideal scenario. That is, it is a measure of the coverage relative to the upper triangle.

The outcome measure was the TIMSS scaled total mathematics score for the country. It was available for grades three, four, seven, and eight.

Results

Country-level regression analyses were performed to examine the curriculum effects as defined by these indicators of coherence and focus. Complete data were available for twenty to thirty-three countries, depending on the grade level.

Before we turn to the regression analyses, it is informative to look at the distribution of two of the main statistics used in these regression analyses. Figure 4-1 illustrates the distribution (by box and whisker plots where the extreme values, or hinges, are the fifth and ninety-fifth percentiles) for the measure of coherence at each of grades three, four, seven, and eight.

The variability across countries is quite small in grades three and four. Remember that the total size of the coherence model at these two grades is thirteen in grade three and twenty-eight in grade four, and, as seen in figure 4-1, the median values are eleven and twenty-three, respectively. However, at grades seven and eight the variability is much larger, with an interquartile range at grade eight of almost twenty. At these grades, no country achieves the maximum values.

Figure 4-1. *Number of Topic-Grade Combinations Aligned with Ideal Scenario of Coherence*

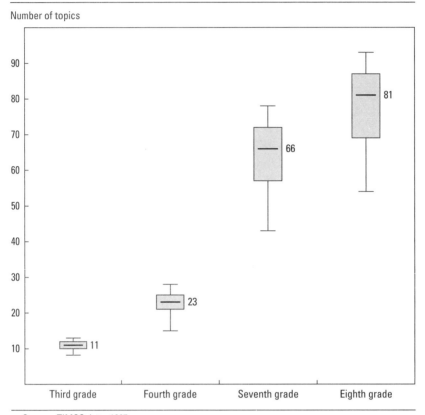

Number of topics

Source: TIMSS data, 1997.

Figure 4-2 represents the distribution of the specific measure of focus across all four grades. No country has a value of zero, indicating that all countries have some topic and grade combinations that occur before those specified in the coherence model. Some countries have ten or fewer premature topic and grade combinations. The median values at grades seven and eight are thirty-three, with some countries having more than sixty such combinations. The variability represented here is quite large, with a difference of as much as fifty topic and grade combinations among some countries at grades seven and eight. The distributions for the total number of topic and grade combinations parallel those in the figure in terms of the large variability across countries and as a result are not presented here. From the point of view of mathematics, the variation portrayed in figure 4-2 is extremely disturbing because it is in this area of the matrix where the hierarchical nature

Figure 4-2. *Number of Topic-Grade Combinations Intended to Be Covered Prematurely*

Number of topics

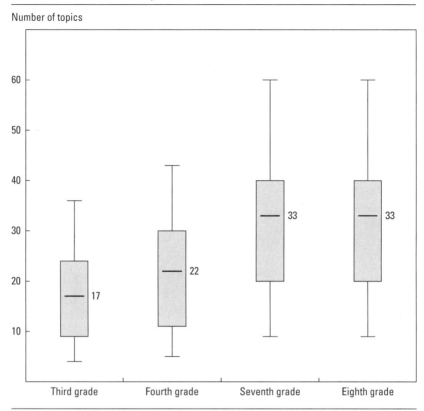

Source: TIMSS data, 1997.

of mathematics can be most violated by the sequencing of school topics in mathematics.

Table 4-2 presents the number of topic and grade combinations in the three categories related to the upper triangle as well as the sum of these—which is one of the focus measures—for a sample of countries at grade eight. Using the data here, one can see that the countries used to develop the model scenario are not necessarily the ones with the greatest degree of consistency with it. Hong Kong, for example, has only about 60 percent agreement with the model of coherence and the Czech Republic about 80 percent. Norway and the United States have a higher value for the coherence measure than any of the six top-achieving countries. The United States also has the most topics introduced too early (63).

One concern we had with this methodological approach was whether using the top-achieving countries to help define the model scenario was "stacking the

Table 4-2. *Alignment with Ideal Scenario of Coherence at Eighth Grade for a Sample of Countries*

Country	Total number of topic/grade combinations in curriculum	Number of topic/grade combinations not aligned (introduced before the ideal scenario)	Number of topic/grade combinations aligned with the ideal scenario	Number of topic/grade combinations not aligned (introduced following the ideal scenario)
Australia	136	32	75	29
Belgium (Fl)	155	37	87	31
Czech Republic	119	32	83	4
France	156	40	85	31
Germany	117	33	74	10
Hong Kong	79	9	62	8
Hungary	158	49	93	16
Japan	128	39	86	3
Korea	125	27	90	8
Netherlands	97	14	54	29
Norway	159	33	96	30
Portugal	144	48	84	12
Singapore	115	11	84	21
Spain	109	23	69	17
United States	186	63	98	25

Source: TIMSS data, 1997.

deck" in terms of the regression analyses we are about to present. Table 4-2 helps to allay those fears. The other factor that helps is that the top-achieving countries chosen were those that were the highest performers at eighth grade. These six countries were not all among the top-achieving countries at grades three, four, and seven. Thus, if there were bias, it would mostly show up at eighth grade. It will become evident that is not the case.

The regression analyses relating the focus measure (the total number of topic and grade combinations intended) and the coherence measure, each separately with country-level achievement, were not statistically significant.[10] Neither focus nor coherence considered alone was systematically related to country-level achievement. This was true at all four grades. Coherence and focus were themselves correlated; figure 4-3 shows the scatter plot at eighth grade of the measure of coherence with the focus measure that deals with early coverage of the topics (r = .77).[11] We present this measure of focus graphically since, as stated previously, it represents the biggest threat to coherence and the largest

Figure 4-3. *Number of Topic-Grade Combinations NOT Aligned (Premature Coverage) versus Number of Topic-Grade Combinations Aligned with Ideal Scenario of Coherence*

Number of topic-grade combinations aligned

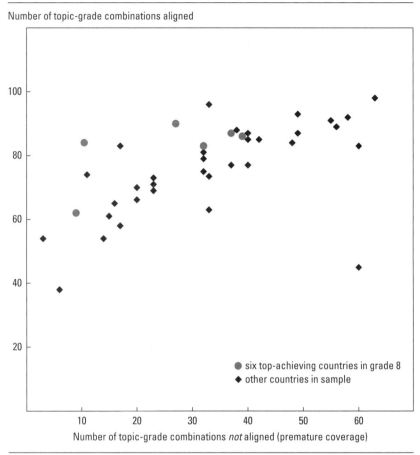

Number of topic-grade combinations *not* aligned (premature coverage)

Source: TIMSS data, 1997.

contribution to the total focus measure (other than the count of the match with coherence).

Let us first consider the results of the regression analyses relating achievement to coherence and focus, with focus being defined as the total number of topics intended up to the grade in question. The coherence measure also is adjusted, as described previously, to each of grades three, four, seven, and eight to correspond to the achievement test results at the same grades. These results are presented in table 4-3.

The significance level of the model fitting varied from $p < .12$ to $p < .01$ across the four grades. At grades three and four, the significance level is greatly

Table 4-3. *Regression Analyses Relating Coherence and Focus to Achievement*

	Third Grade			Fourth Grade			Seventh Grade			Eighth Grade		
Predictor	Estimate	Std. error	p	Estimate	Std. error	p	Estimate	Std. error	p	Estimate	Std. error	p
Model												
Number of topic/grade combinations aligned with ideal scenario	9.94	7.38	0.196	7.48	4.04	0.081	3.57	1.48	0.023	2.83	1.12	0.017
Total number of topic/grade combinations in curriculum	-2.42	1.13	0.047	-2.58	1.05	0.025	-1.64	0.57	0.008	-1.39	0.45	0.004
Model fit												
R-Squared	0.2191			0.2574			0.2205			0.249		
Residual mean squares	1652.9			1489.0			1476.0			1331.4		
p<	0.1222			0.0687			0.0270			0.0136		
Number of countries	20			21			32			33		
Standardized coefficient												
Number of topic/grade combinations aligned with ideal scenario	0.4374			0.7331			0.8481			0.8240		
Total number of topic/grade combinations in curriculum	-0.6969			-0.9703			-1.0044			-1.0204		

Source: TIMSS data, 1997.

influenced by the small sample sizes of twenty and twenty-one countries (fewer countries in TIMSS participated at the lower grades). The estimated R^2 values indicated that the two indicators accounted for around 20 to 25 percent of the variance. In all four instances, the focus measure was statistically significant, indicating that, when controlling for coherence, the larger the number of topic and grade combinations intended for coverage, the lower the mean country achievement.

The coherence measure was significant ($p < .02$) at grades seven and eight and only marginally so at the fourth grade ($p < .08$). Here the coefficients are all positive, indicating that the greater the degree of match of a country's intended curriculum to the ideal scenario, the higher the country's mean performance on the TIMSS test. The lack of significance at grade three is probably a combination of two factors. One is the small sample size. The other, and likely more important, reason is the small size of the ideal scenario (only thirteen cells) and the corresponding lack of much variability at grade three for that scenario.

The results of these analyses at eighth grade imply that a country could raise its mathematics achievement by about one-fourth of a standard deviation if it increased its coverage of the ideal scenario by ten topic and grade combinations or intended opportunities to learn, while holding constant the current level of focus. (The increase at seventh grade would be more than a third of a standard deviation; the predicted increase at grades three and four is even greater, but the precision of these estimates is marginal at best.)

For the focus measure (where the variation is much larger, as displayed in figure 4-2), a decrease of fifty in the number of intended topic and grade combinations would predict an increase in achievement of almost three-fourths of a standard deviation. This estimate is especially relevant to the United States, which has the highest number of topic and grade combinations and where a decrease of fifty combinations would bring it more into line with Japan and Korea (see table 4-2). The predicted effect for focus would be even greater at grades three, four, and seven. For both coherence and focus, ignoring the marginal nature of the significance at grade four, the implication seems to be that the predicted effects are greater at the earlier grades. Perhaps premature coverage of topics has a more deleterious effect at the earlier grades. This, of course, is only a hypothesis in need of further study.

The next set of analyses used the same measure of coherence but included the special focus measure (table 4-4). Here the significance level of the model fitting varied similarly to the other model fitting (from $p < .12$ to $p < .03$). The same was true of the similarities in the estimated R^2. The lack of focus resulting from the premature coverage (at least from the point of view of the model scenario) of topics at earlier grades was statistically significant at each grade level. The estimated coefficients were all negative as well.

Table 4-4. Regression Analyses Relating Coherence and Specific Focus to Achievement

Predictor	Third Grade			Fourth Grade			Seventh Grade			Eighth Grade		
	Estimate	Std. error	p	Estimate	Std. erorr	p	Estimate	Std. error	p	Estimate	Std. error	p
Model												
Number of topic/grade combinations aligned with ideal scenario	7.52	6.57	0.268	4.91	3.19	0.141	1.84	1.06	0.094	1.66	0.87	0.068
Total number of topic/grade combinations in curriculum	-2.42	1.13	0.047	-2.58	1.05	0.025	-1.79	0.71	0.017	-1.97	0.70	0.009
Model fit												
R-Squared	0.2191			0.2574			0.1826			0.2122		
Residual mean squares	1652.9			1489.0			1547.9			1396.6		
p<	0.1222			0.0687			0.0538			0.0279		
Number of countries	20			21			32			33		
Standardized coefficient												
Number of topic/grade combinations aligned with ideal scenario	0.3309			0.4807			0.4382			0.4825		
Total number of topic/grade combinations not aligned (premature coverage)	-0.6209			-0.7646			-0.6405			-0.7151		

Source: TIMSS data, 1997.

Conditioning on this measure of focus, coherence was not significant at either the third or fourth grade. For the seventh and eighth grades, however, the coherence measure was significant with estimated positive coefficients.

The final set of analyses included the ratio that combined total coverage (focus) with degree of overlap with the ideal scenario. The model varied from marginally significant at grade three ($p < .10$) to statistically significant at the other three grades. In all cases the estimated regression coefficients were positive, indicating that if a country spent more of its total effort (defined as the combination of all cells for which coverage was intended) on the topic and grade combinations related to the model scenario, then higher mean student achievement would occur at each of grades four, seven, and eight. The scatter plot for grade eight is presented in figure 4-4.

Discussion

We have contended elsewhere that coherence is a critical, if not the single most important, element defining high-quality content standards. Such content standards are defined in terms of the disciplines that underlie the school subject matter articulated in those standards. We suggest that standards are coherent if they specify topics, including the depth at which the topic is to be studied as well as the sequencing of the topics, within each grade and across grades, in a way that is consistent with the structure of the underlying discipline.[12]

The results of the above analyses are consistent with that contention—especially at the upper grades, where we found the measure of coherence to be positively related to achievement. There is also some indication that coherent standards are at least marginally related to student performance at fourth grade as well. The inconsistency of the statistical significance ($p < .05$) of the relative measure (coherence over focus) with the marginal significance of the other analyses ($p < .07$) at fourth grade warrants further investigation, especially with a larger number of countries.

The other striking result is that even when controlling for coherence, the focus measure was statistically significant at all grades. Even at third grade, where the overall model was not statistically significant, the focus measure was significant ($p < .05$). This finding implies that for a country to have a high mean level of performance, it must have a high degree of focus as well as coherence. The amount of "clutter" created by covering too many topics too early or before their time from a mathematical point of view must be kept small. Covering too many topics does have a negative impact on student learning even when controlling for coherence.

As further evidence of the effect of the interplay between coherence and focus on achievement, the bivariate relationship between focus and achieve-

Figure 4-4. *Eighth Grade Mathematics International Scaled Score versus the Ratio of Coherence to Focus*

Eighth grade math scaled score

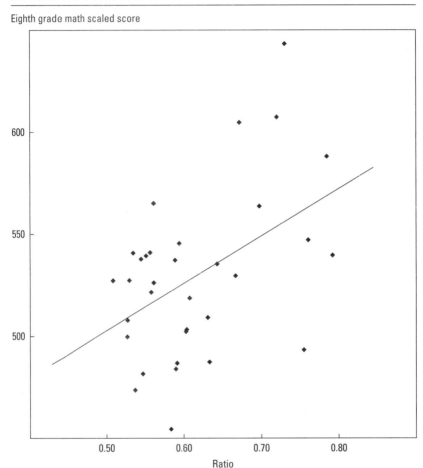

Ratio

Source: TIMSS data, 1997.

ment is not statistically significant at any of the four grade levels. In other words, focus by itself is not related to country-level achievement, but it is related to achievement (as reported above) only when coherence is included in the model. This implies that focus only has a statistically significant negative impact conditioned on coherence. Put simply, no matter what the degree of coherence evident for a country, covering topics at grades outside those defining coherence has a negative impact.

The lack of statistical significance for the focus measure in predicting country-level achievement is consistent with the findings of Baker and

LeTendre. Baker, as a discussant for this paper, drew the conclusion that such measures are not related to achievement and as a result are not important for characterizing cross-national differences in curriculum. This conclusion supports his general point that cross-national differences in curriculum have mostly disappeared with the increasing globalization of curriculum. However, we have concluded the opposite, not only here but in other reports as well.[13]

Measures of curriculum that are not embedded in the substance of the subject matter—here mathematics—and that measure more generalized (and thus, in some ways, superficial) aspects of subject matter are not adequate descriptors of curriculum. Focus is a good example. It is not a significant predictor of achievement by itself—not in Baker's analyses, not in our analyses. When combined with a measure that does derive directly from the subject matter, such as coherence, focus does predict achievement. The measure of coherence can only be defined in terms of the substance of the particular subject matter. Its definition would be different for other subject matters. Measures of focus, breadth, or depth do not have this characteristic—they are essentially the same regardless of subject matter. Types of measures derived *from the content itself* do characterize important cross-national differences that are related to cross-national differences in achievement. Content, at least in mathematics, has yet to become so globalized and homogeneous in substance and sequence that there are not important cross-national differences to be identified and related to achievement differences.

The policy implications for the United States are profound. When we examined the mathematics national professional standards developed by the NCTM in other work, the contrast with the ideal scenario was striking.[14] Table 4-5 graphically represents the NCTM standards that were in effect at the time of TIMSS. This contrast is equally striking when we more recently examined the standards of more than 100 local districts and the individual standards in more than forty states.

Not only is the organizing principle underlying these curricula unlike that of the ideal scenario, but the organizing principle for these U.S. standards seems to be *qualitatively* different rather than simply *quantitatively* different (differing in degree). Movement toward more coherent standards has most recently become available with the new NCTM focal point standards.[15] They are intended to focus on topics more in line with the principles of coherence. Table 4-6 shows these new standards against the backdrop of the coherence model.

A methodological caveat needs to be made here about using the results of these analyses in making U.S. policy. Statistically these results are appropriate only at the country level (the level at which the analyses were done) and as such do not speak directly to what would happen if such an analysis were done at

Table 4-5. *Mathematics Topics Intended at Each Grade by the 1989 NCTM Standards*

Topic	1	2	3	4	5	6	7	8
Whole number: meaning	•	•	•	•	•	•	•	•
Whole number: operations	•	•	•	•	•	•	•	•
Measurement units	•	•	•	•	•	•	•	•
Common fractions	•	•	•	•	•	•	•	•
Equations & formulas	•	•	•	•	•	•	•	•
Data representation & analysis	•	•	•	•	•	•	•	•
2-D geometry: basics	•	•	•	•	•	•	•	•
2-D geometry: polygons & circles	•	•	•	•	•	•	•	•
Measurement: perimeter, area, & volume	•	•	•	•	•	•	•	•
Rounding & significant figures	•	•	•	•	•	•	•	•
Estimating computations	•	•	•	•	•	•	•	•
Whole numbers: properties of operations	•	•	•	•	•	•	•	•
Estimating quantity & size	•	•	•	•	•	•	•	•
Decimal fractions	•	•	•	•	•	•	•	•
Relation of common & decimal fractions	•	•	•	•	•	•	•	•
Properties of common & decimal fractions					•	•	•	•
Percentages					•	•	•	•
Proportionality concepts					•	•	•	•
Proportionality problems					•	•	•	•
2-D geometry: coordinate geometry	•	•	•	•	•	•	•	•
Geometry: transformations	•	•	•	•	•	•	•	•
Negative numbers, integers, & their properties					•	•	•	•
Number theory	•	•	•	•	•	•	•	•
Exponents, roots, & radicals					•	•	•	•
Exponents & orders of magnitude	•	•	•	•	•	•	•	•
Measurement: estimation & errors	•	•	•	•	•	•	•	•
Constructions using straightedge & compass	•	•	•	•	•	•	•	•
3-D geometry	•	•	•	•	•	•	•	•
Geometry: congruence & similarity	•	•	•	•	•	•	•	•
Rational numbers & their properties					•	•	•	•
Patterns, relations, & functions	•	•	•	•	•	•	•	•
Proportionality: slope & trigonometry					•	•	•	•

Source: TIMSS data, 1997.
Top-achieving countries topic profile.
• 1989 NCTM Standards Intended Topics.

Table 4-6. *Mathematics Topics Intended at Each Grade by the 2006 NCTM Focal Points*

Topic	Grade							
	1	2	3	4	5	6	7	8
Whole numbers: meaning	•	•	•	•	•	•		
Whole numbers: operations	•	•	•	•	•			
Measurement: units	•	•	•	•	•	•	•	•
Common fractions			•	•	•	•	•	
Equations & formulas			•	•	•		•	•
Data representation & analysis	•		•	•	•		•	•
2-D geometry: basics			•	•				•
2-D geometry: polygons & circles	•	•	•	•	•		•	•
Perimeter, area, & volume			•	•	•	•	•	•
Rounding & significant figures								
Estimating computations		•	•	•	•			
Properties of operations	•	•	•	•	•			
Estimating quantity & size			•					
Decimal fractions			•		•	•		
Relationships of common & decimal fractions		•	•			•	•	
Properties of common & decimal fractions					•	•		
Percentages							•	
Proportionality concepts		•				•	•	•
Proportionality problems						•	•	•
2-D coordinate geometry					•			•
Transformations	•		•	•				
Negative numbers, integers, & their properties					•		•	
Number theory	•				•		•	
Exponents, roots, & radicals								•
Exponents & orders of magnitude								•
Estimation & errors		•	•		•			
Constructions w/ straightedge & compass								
3-D geometry	•	•			•		•	•
Congruence & similarity	•		•				•	
Rational numbers & their properties						•		
Patterns, relations, & functions	•	•	•	•	•	•		•
Slope & trigonometry							•	•

Source: TIMSS data, 1997.
Top-Achieving Countries.
• 2000 NCTM Focal Points.

state or district levels within the United States. However, the results do seem to have implications that warrant further investigation.

The results of these analyses do not bode were for the United States. They predict poor levels of performance, especially in the middle grades. In fact, poor performance in the middle grades has been seen repeatedly, as recently as 2003.

Notes

1. W. H. Schmidt and others, *Many Visions, Many Aims,* vol. 1: *A Cross-National Investigation of Curricular Intentions in School Mathematics* (Dordrecht, Netherlands: Kluwer, 1997); W. H. Schmidt, C. McKnight, and S. Raizen, *A Splintered Vision: An Investigation of U.S. Science and Mathematics Education* (Dordrecht, Netherlands: Kluwer, 1997).
2. W. H. Schmidt and others, *Why Schools Matter: A Cross-National Comparison of Curriculum and Learning* (San Francisco: Jossey-Bass, 2001).
3. W. H. Schmidt, H. A. Wang, and C. C. McKnight, "Curriculum Coherence: An Examination of US Mathematics and Science Content Standards from an International Perspective," *Journal of Curriculum Studies* 37, no. 5 (2005): 525–59.
4. Schmidt, Wang, and McKnight, "Curriculum Coherence," citing J. S. Bruner, "On Learning Mathematics," *Mathematics Teacher* 88, no. 4 (1995): 330–35.
5. D. P. Baker and G. K. LeTendre, *National Differences, Global Similarities: World Culture and the Future of Schooling* (Stanford, Calif.: Stanford Social Sciences, 2005), p. 162.
6. For a characterization of the TIMSS for 1995, see Schmidt and others, *Why Schools Matter;* W. H. Schmidt and C. C. McKnight, "Surveying Educational Opportunity in Mathematics and Science: An International Perspective," *Educational Evaluation and Policy Analysis* 17, no. 3 (1995): 337–53; and A. E. Beaton and others, *Mathematics Achievement in the Middle School Years: IEA's Third International Mathematics and Science Study* (Boston College, 1995).
7. For a description of the framework and coding procedures, see Schmidt and others, *Many Visions, Many Aims.*
8. Schmidt, Wang, and McKnight, "Curriculum Coherence," pp. 532–33.
9. Ibid.
10. The significance levels were as follows: grade 3 ($p < .11$); grade 4 ($p < .13$); grade 7 ($p < .16$) and grade 8 ($p < .09$).
11. The estimated correlation coefficients for the other grades are: grade three ($r = .67$); grade four ($r = .76$) and grade seven ($r = .75$).
12. Schmidt, Wang, and McKnight, "Curriculum Coherence."
13. Schmidt and others, *Many Visions, Many Aims;* Schmidt and others, *Why Schools Matter;* and W. H. Schmidt and others, *Facing the Consequences: Using TIMSS for a Closer Look at U.S. Mathematics and Science Education* (Dordrecht, Netherlands: Kluwer, 1999).
14. Schmidt, Wang, and McKnight, "Curriculum Coherence."
15. National Council of Teachers of Mathematics, *Curriculum Focal Points for Prekindergarten through Grade 8 Mathematics: A Quest for Coherence* (Reston, Va.: National Council of Teachers of Mathematics, 2006).

5

U.S. Algebra Performance in an International Context

JEREMY KILPATRICK, VILMA MESA, AND FINBARR SLOANE

In this chapter we use data on how students performed on mathematics items in the Trends in International Mathematics and Science Study (TIMSS) to explore some effects of the setting in which U.S. students learn algebra. We begin by describing characteristics that make the U.S. school mathematics curriculum fundamentally different from that of other countries. We locate these characteristics in their historical context, contrast them with visions of mathematics and mathematics learning in other countries, and then, drawing on data from TIMSS, use the bulk of the chapter to examine in detail the performance of U.S. students relative to that of their peers elsewhere. We look at the content of the TIMSS algebra items, the representations they contain, and the cognitive demands they make, with particular attention to those items on which U.S. students have done especially well or especially poorly in either an absolute or a relative sense. The profile of U.S. students' performance on these items suggests they have encountered a school mathematics curriculum that treats algebra as generalized arithmetic only, rather than one that adopts a function-oriented approach common in other countries and recently advocated in the United States.[1]

We are grateful to Gerald LeTendre for comments on an early draft of this chapter; to Joseph Galia, Michael Martin, and Ina Mullis for assistance in obtaining and interpreting TIMSS data; and to Charalambos Charalambous and Erica Boehne for assistance with the data analysis.

The college-preparatory high school curriculum in the United States is virtually unique in devoting two or three yearlong courses to algebra, with geometry given a separate year of its own. In other countries, students are taught algebra and geometry (and other areas of mathematics) simultaneously, in either integrated mathematics courses or parallel strands.[2] William Schmidt describes the difference: "Better U.S. mathematics students during high school years take separate courses in geometry, pre-calculus, etc. In most TIMSS countries, students take a course in mathematics—a course which may include studying parts of advanced algebra, geometry, finite mathematics, and calculus at the same time. They may take such courses for several years."[3] In international curricula, therefore, mathematics is seen more as an integrated whole than as subject matter that can be parsed by topic area without overt integration. This curricular parsing is the same for weak and strong students in the United States. The integration is left up to the student, a task many find difficult or fail to complete on their own.

Another distinguishing feature of the U.S. curriculum is that, until recently, school algebra and geometry courses were seen as reserved for the elite—those heading for college—while everyone else was offered another set of courses, labeled with such terms as "general mathematics," in which almost no attention was given to algebra or geometry. Consequently, Americans typically have seen algebra as a subject beyond the capacity of the average high school student.

In this paper, after a brief examination of sources of the nature and structure of the U.S. algebra curriculum, we describe U.S. students' performance in algebra through an analysis of TIMSS items from 1995, 1999, and 2003. TIMSS is a fruitful site to use for investigation for several reasons. First, the achievement test items in TIMSS have been designed to assess knowledge associated with the school mathematics curriculum, rather than to measure students' mathematical literacy, as in the Program for International Student Assessment (PISA).[4] Second, TIMSS has refined the items used to assess knowledge of algebra (and other topics) to reflect changes in curricula and the way mathematics is taught.[5] Finally, TIMSS collects data on a large set of countries, which enables researchers to study the relative performance of students who have learned algebra in different ways and under different conditions.[6] Of all the TIMSS content topics in science or mathematics, algebra is the only one in which there was at least a moderate association across countries between topic coverage in the curriculum and the students' performance on the topic.[7]

We look at TIMSS items on which U.S. students did well and did poorly, both in absolute terms and in comparison with students in other countries. Our description of their performance takes into account several dimensions of the tasks, provides a picture of those aspects of algebra in which U.S. students, as a whole, are strong or weak, and compares their performance with that of students in other countries. That description, by revealing patterns of performance, can suggest areas that might be the focus of curricular and instructional attention.

History of U.S. Algebra Courses

The layered course arrangement with tracking is a product of the history of college-entrance requirements in the United States and has remained in place owing, in part, to the absence of a national curriculum authority that might have mandated a different arrangement. Arithmetic and geometry entered the Harvard curriculum during the seventeenth century and by the late eighteenth century had become lower-division (freshman or sophomore) courses, whereas algebra, a relative newcomer, was being offered in the senior year.[8] In 1787 Harvard reorganized its curriculum to put arithmetic in the freshman year and "algebra and other branches of mathematics" in the sophomore year.[9] U.S. universities started requiring arithmetic for admission, and some time before 1820 Harvard began to require algebra. Other universities soon followed suit. Not until after the Civil War, however, was geometry required for entrance. The order in which these mathematical subjects were first required for college entrance shaped, and continues to shape, the college-preparatory courses offered in secondary school, with each new requirement acting like a geological stratum laid down by natural forces, thus pressing the courses required earlier into lower grades.

Despite repeated efforts to combine algebra, geometry, and other mathematics topics into integrated mathematics courses, the U.S. first course in algebra—like the yearlong course in geometry—has kept its title and much of its form.[10] Positioned for more than a century in the ninth grade, Algebra I recently began to move into the eighth grade, partly as a consequence of research showing that other countries were offering more algebra earlier than the United States was.[11]

Nonetheless, it is still the case that although algebra and geometry are important elements of the middle school curriculum in other TIMSS countries, U.S. middle schools tend to offer these topics to relatively few students. As Edward Silver commented in 1998, "Compared to many other countries, the content taught at grade 8 in the United States is similar to the content taught at grade 7 elsewhere, and the performance expectations are lower in the United States."[12] The proportion of eighth graders taking the National Assessment of Educational Progress (NAEP) mathematics assessment who reported taking algebra rose from 20 percent in 1992 to 25 percent in 1996.[13] In 2000, 31 percent of eighth graders reported taking pre-algebra, and 26 percent reported taking algebra.[14]

Although the number of students taking advanced high school mathematics courses remains a minority, enrollments in such courses have grown rapidly.

From 1982 to 1998, the proportion of students taking mathematics courses beyond geometry and algebra II increased from 26 percent of all high school graduates to 41 percent. By 2000 it was 45 percent.[15] Such growth suggests that the number of students beginning the study of algebra in eighth grade is likely to continue to increase.

History of U.S. Approaches to School Algebra

School algebra in the United States has taken on a variety of forms even as it has remained enshrined in separate yearlong courses. When it entered the high school curriculum, algebra was largely an extension and generalization of school arithmetic. It was built up by induction on a base of numerical quantities and operations on them. As the nineteenth century drew to a close, school algebra, like other subjects, came under the influence of so-called faculty psychology, which saw the mind as having separate faculties or powers. To exercise those powers, students needed to drill and practice their algebraic skills, and teachers and textbook writers were happy to provide ample opportunity for such activity. School algebra came to be seen as a purely mathematical discipline, with little or no practical value.

Meanwhile, other countries, especially those in Europe, were being influenced by developments in mathematics, and particularly by the proposal of the German mathematician Felix Klein to use the function concept as a basis for developing not simply algebra but the entire secondary mathematics curriculum. The function concept had some advocates in the United States, but it was very much a minor influence. Algebra as generalized arithmetic was the mainstream approach until the new math movement began to influence the U.S. curriculum in the early 1960s. Algebra was only then recast as a more abstract subject, with functions playing a central role and with definitions based on set theory.

Recent developments in the U.S. school mathematics curricula suggest that an approach to algebra as the study and use of functions rather than as simply equation solving and manipulation of expressions may be gaining ground.[16] The approach, however, is different from the set-theoretical approach taken during the new math era. Functions are introduced as rules for connecting one number to another, with equations used to model sets of data for the purpose of solving problems. This new approach makes heavy use of technology, especially to capitalize on its ability to manipulate linked tabular, symbolic, and graphical representations of functions. Students can work with complicated equations that would be difficult, if not impossible, to represent or solve using paper and pencil alone. This modeling approach to algebra, although growing, is still very much a minor theme in U.S. school mathematics.[17]

U.S. Students' Performance in TIMSS 2003

Before taking a closer look at how U.S. students performed on algebra items in the three TIMSS assessments, we first consider how they did on items from various categories of mathematics content in TIMSS 2003. We consider grades four and eight only.

We should note that the TIMSS mathematics frameworks and specifications were changed between TIMSS 1999 and TIMSS 2003.[18] For example, the 1995 category *Patterns, relations, and functions,* which served as a surrogate for algebra at grade four, was changed to *Patterns, equations, and relationships* in 2003 and given half again as much weight in the mix of items. At grade eight, the category name remained "Algebra," but it was given almost twice as much weight in the 2003 assessment. The total number of items was increased for both grades between 1995 and 2003. Because of these and other changes, we do not give details regarding U.S. students' performance in 1995 or 1999.[19]

For 2003 the data for each content category were put on the same TIMSS scale using Item Response Theory (IRT) equating. This scaling allows a more precise determination of how U.S. students' performance differs from that of students in other countries than the more intuitively clear average number of items correct. In this section, we use the scale adopted by TIMSS for reporting results; it has a mean of 500 and a standard deviation of 100.[20]

Grade Four Scores

In TIMSS 2003 the average scaled score of the twenty-five countries participating in the analysis of mathematics performance was 495. U.S. fourth-grade students were above the international average for every topic and for all items taken together.[21] Overall, they did not show any change in performance between 1995 and 2003 as measured by the TIMSS scale. There were too few items common to both assessments, however, to test the significance of performance changes in each content category. Table 5-1 shows the number of items in the five mathematics content categories, the scaled score the average U.S. fourth grader received on the items in that category, and the average scaled score across all twenty-five TIMSS countries.

The performance of U.S. fourth graders on items dealing with *Data* (data handling and probability) was relatively strong—as it had been in 1995. *Measurement* remained the one area of relative weakness, the only one where the U.S. fourth graders did not score significantly above the international average. And again, performance on *Patterns, equations, and relationships* items indicated that U.S. fourth graders were above average in algebra compared with students in other countries. Only six of the twenty-five countries outperformed the U.S. fourth graders in *Patterns, equations, and relationships,* and only four countries

Table 5-1. *Performance of U.S. and All TIMSS 2003 Fourth Graders by Content Category*

Content category	Number of items	U.S. average scaled score	25-country average scaled score
Number	63	516	495
Patterns, equations, and relationships	22	524	495
Measurement	33	500	495
Geometry	24	518	495
Data	17	549	495
Total	161	518	495

Source: Data from Rodger W. Bybee and others, "TIMSS 2003: An Introduction," *The Natural Selection: The Journal of BSCS* (Winter 2005): 8, and Michael O. Martin, Ina V. S. Mullis, and Steven J. Chrostowski, eds., *TIMSS 2003 Technical Report* (Boston College, 2004), appendix E.

in *Data*. Even though the framework and the distribution of items changed somewhat across the TIMSS cycles, "the pattern of performance is similar to what it was in 1995. There are some hints in the results that U.S. fourth graders may be doing a bit better in patterns, equations, and relationships than in 1995, but not as well in geometry."[22]

Grade Eight Scores

The results of the TIMSS 2003 assessment show that, overall, U.S. eighth graders exceeded the international average (466) of the forty-eight countries in mathematics performance.[23] Between 1995 and 2003, U.S. eighth graders showed significant improvement in their average performance relative to that of other TIMSS countries. Table 5-2 shows the number of items in the five content cat-

Table 5-2. *Performance of U.S. and All TIMSS 2003 Eighth Graders by Content Category*

Content category	Number of items	U.S. average scaled score	48-country average scaled score
Number	57	508	466
Algebra	47	510	466
Measurement	31	495	466
Geometry	31	472	466
Data	28	527	466
Total	194	504	466

Source: Data from Rodger W. Bybee and others, "TIMSS 2003: An Introduction," p. 9, and Martin, Mullis, and Chrostowski, *TIMSS 2003 Technical Report,* appendix E.

egories and the scaled score on those items for the average U.S. eighth grader compared with the average across the TIMSS countries.

In 2003 U.S. eighth graders were above the international average in every category of mathematics. As in 1995, they did relatively best on the items dealing with *Data* and relatively poorly in *Measurement* and *Geometry*, but in 2003, their poorest performance was in *Geometry* and not *Measurement*. In only eight of the forty-eight countries did eighth graders outperform the U.S. eighth graders in *Algebra* and *Data*, "a pattern similar to that of the U.S. fourth graders in 2003."[24] From 1999 to 2003, U.S. eighth graders gained significantly, on average, in their performance on TIMSS items classified as dealing with *Algebra* or with *Data*.[25]

U.S. Students' Performance in Algebra

The evidence just presented, coupled with data for other national and international assessments, indicates that in algebra, U.S. students perform at roughly the same level as, or somewhat better than, they do across all domains of mathematics. The evidence from TIMSS 1995, 1999, and 2003 suggests as well that their performance in algebra is about average or somewhat above compared with that of students in other countries. Their performance in algebra contrasts with their performance in geometry and measurement, which has tended to be weaker than that of students in other countries, and in probability and statistics, which has tended to be stronger.

U.S. students' average level of performance in algebra, however, masks considerable variation across items and across students, the latter in large part because of the extensive tracking already noted. Some U.S. eighth graders, for example, are taking a full year of algebra, whereas others are still studying arithmetic.[26] Below we report the results of analyses of TIMSS data that attempt to uncover and disentangle some facets of U.S. students' performance in algebra.

Item Characteristics

We chose to center our analyses on characteristics of the items. Following a practice that other investigators have also used, we asked what was common across the items on which U.S. students performed in the first or fourth quartile.[27] In other words, what are the characteristics of the items to which 25 percent or fewer of the U.S. students responded correctly, and what are the characteristics of the items to which 75 percent or more of the U.S. students responded correctly? This approach provides information about the kinds of algebraic work on which U.S. students appear to be most and least proficient. We also asked about characteristics of the items on which U.S.

students outperformed or underperformed students from other countries. By pursuing these analyses, we aimed at describing strengths and weaknesses of U.S. student performance that would illuminate the nature of the attained curriculum and would offer suggestions as to what should be maintained and what should be modified.

Because most of our discussion is based on item data, we needed to identify those TIMSS items related to algebra and algebraic thinking. We examined all the fourth-grade items classified by TIMSS as *Patterns, relations, and functions* in 1995 or *Patterns, equations, and relationships* in 2003; all the eighth-grade items classified as *Algebra* in 1995, 1999, or 2003; and all the twelfth-grade mathematics literacy items in 1995. To more fully characterize the items, we chose three dimensions of the items: algebraic content, representation use, and cognitive demand.

Algebraic Content. The main purpose of the content classification was to respond to the question: Algebraically speaking, what is this item about? We began with a classification scheme developed for an analysis of items from a special NAEP study of eighth graders who were taking or had taken algebra.[28] Because many of the TIMSS algebra items required students to carry out or recognize algebraic manipulations, we added the category "algebraic manipulation" to the classification scheme. Also, through the process of categorizing items, we revised the definitions of the categories to provide a more reliable system for content classification, as follows:

—*Patterns:* Find the next several numbers in a given pattern and write the rule; or extend a pattern of sums and choose the correct answer.

—*Informal algebra:* Select the solution to an equation presented pictorially; model an arithmetic expression, possibly generated in a real-world scenario; or choose a value that satisfies an arithmetic relationship.

—*Equations:* Solve an equation or a system of linear equations; solve a linear equation in two variables for one variable in terms of the other; or determine which equation in two variables fits the values given in a table.

—*Inequalities:* Choose a value that satisfies a simple linear inequality.

—*Functions:* From a verbal description, select an expression for a linear function described in a realistic context; identify a rule relating two variables or sets of numbers; or using a graphical representation, deduce the value of one variable given the other.

—*Algebraic reasoning:* Given information about pairwise relationships between quantities, draw a conclusion about their order; or given a linear function, draw a conclusion about the effect of a change in either x or $f(x)$ on the other.

—*Algebraic manipulation:* Transform an expression by means of symbolic manipulation; identify an expression that corresponds to a verbal statement; or interpret a variable.

We emphasize that some of these categories can be seen as equivalent mathematically; for example, writing a rule for a pattern (*Patterns*) and identifying a rule relating two variables (*Functions*) can be interpreted as essentially the same mathematical activity. Research suggests, however, that from the student's point of view, the two activities are different. Students' knowledge becomes compartmentalized in ways that render equivalent activities separate and sometimes incompatible.[29] Similarly, we contrast *Functions* with *Algebraic reasoning* as follows: A functions item requires students either to identify the function relating two variables or to calculate the value of one variable given the other, whereas an algebraic reasoning item requires students to reason about a relation among variables that may or may not be a function.

Representation Use. To describe the type of representation entailed in an item, we used the following categories: numerical, verbal, graphical, symbolic, and pictorial (with N, V, G, S, and P, respectively, as codes).[30] Each item received two codes: one to identify the representation used to present the item, and a second to identify the representation needed for the response. For multiple-choice items, the choices were used to establish the representation sought. For constructed-response items, we analyzed the rubrics to see which types of representation were considered correct.

Cognitive Demand. In addition to using content and representation characteristics, we classified items in terms of cognitive demand using the notion of content-process space.[31] Originally designed to classify assessment items in science, this framework attempts to describe knowledge that is meaningfully organized—that is, knowledge that employs four cognitive activities: problem representation (using underlying principles and relevant concepts), strategy use (efficient, informative, and goal oriented), self-monitoring (ongoing and flexible), and explanation (principled and coherent).[32] Although every task in an assessment does not necessarily have all four characteristics, an examination of the affordances (action possibilities) of each task allows one to determine how demanding it is in terms of the quality of the content addressed and the type of processes required.

In the content-process space, tasks can be classified along two dimensions characterized by their poles: *rich* versus *lean* content, and an *open* versus *constrained* process. The location of the task "is related to the nature and extent of cognitive activity underlying performance," and therefore the space provides a good device for describing cognitive task demand.[33] Figure 5-1 shows the space and defines four quadrants, each of which helps in characterizing the demand. Although it is reasonable to assume that, in general, students would be more likely to do well on items in Quadrant III than on those in Quadrants I or II, the pattern of performance across the four quadrants may be useful in revealing some of the strengths and weaknesses of the curriculum.

Figure 5-1. *Content–Process Space*

		Content dimension
		Rich
	Quadrant II: Activities that are rich in content but more process constrained—tend to measure specific content in particular settings	Quadrant I: Activities that are rich in content and open in process—tend to be complex open-ended tasks that require extensive mobilization of information
Process dimension	*Constrained*	*Open*
	Quadrant III: Activities that are lean in content and process constrained—tend to assess factual knowledge	Quadrant IV: Activities that are lean in content with a more open process—tend to assess originality or creativity
		Lean

Source: Adapted from Gail P. Baxter and Robert Glaser, Investigating the Cognitive Complexity of Science Assessments," *Educational Measurement* 17, no. 3 (1998): 37–45.

Item Coding

For each item, Mesa produced a description of what the item required the student to do, which Kilpatrick and Sloane then checked. Mesa coded the algebraic content and the representations for all the algebra items in 1995, 1999, and 2003 (81 released items). This exercise led to a refinement of the coding system to distinguish between *equation* and *function,* with the former items referring to developing or solving an equation and the latter to expressing or using the dependence relationship between variables. Coding cognitive demand required the production of the solution paths that students might take and an analysis of the coding rubric to decide what was expected from students. This last coding was conducted only with the high- and low-performance items (forty-eight items, or 59 percent of the original sample).

To test the reliability of the coding, a graduate student at the University of Michigan coded eight randomly selected items (10 percent of the total sample). The results showed low reliability for classifying the content of the items, which led to a refinement of the descriptions for the categories *informal algebra* and *equations.* The interrater reliability for coding the representations was high (Cohen's kappa = 80 percent); for the cognitive demand, it was moderate (Cohen's kappa = 67 percent). Examining the rationale for the coding of cognitive demand, we found three items for which the coder had listed more than one way to solve the problem, which under our definition would qualify as *open process*. Similarly, we found two items for which the coder had indicated that a student would have major difficulties as suggested by the literature. In addition, the content for these two items would require instruction before they could be solved. Both of these are major defining characteristics for rich content. Recoding the items to fit our definitions of process and content, we found that Cohen's kappa was 83 percent for process and 71 percent for content, both acceptable measures of interrater reliability.[34]

TIMSS almanac information from each cycle and at each grade was used to create a data file for the following variables: year of testing, grade, item ID, whether the item had been released to the public, country, number of students tested, percentage correct, and standard error.[35] For constructed-response items, the percent correct included the proportion of students who received partial credit. Because some items contained more than one question, we followed the TIMSS practice and considered each question as a separate "item."

Any examination of individual items can be problematic because a single item does not yield enough information to permit inferences about any specific construct that a test purports to measure. By looking at groups of items sharing a common characteristic, however, one can more confidently develop hypotheses to account for students' performance.

Item Comparison

Below we discuss, for each TIMSS cycle, the content and representation of the released items for which performance was high, followed by their cognitive demand. We look at high performance in both an absolute and a relative sense. We then do the same for those released items for which performance was low. We tried as best we could to categorize and analyze the unreleased TIMSS items using item descriptions alone to see if performance on them followed the same patterns as on the released items, but we eventually decided that the results were too speculative to report. Therefore, the analyses described here draw on the released items only.[36] In each section, we present these analyses in the chronological order of the TIMSS cycles (1995, 1999, and 2003).

High-Performance Items

We first consider just those algebra items that at least 75 percent of U.S. students answered correctly, quite apart from how students in other countries did on the items.[37] These items indicate what we call absolute high performance. Even when U.S. students showed high levels of performance, however, students in other countries often outperformed them, so we then look at those items on which U.S. students performed well relative to students in other countries: relative high performance.

Absolute High Performance (p ≥ .75)

Across the three cycles of TIMSS testing, we identified thirteen released items (out of eighty-one classified as algebra-related) and twelve secure items (out of seventy-six classified as algebra-related) that were answered correctly by 75 percent or more of the U.S. students in the sample. Table 5-3 shows the distribution of items across cycles and grades.

1995 Items: Content and Representation. The one released fourth-grade item, Item L04, that was classified as high performance in 1995 asks the student to identify a sequence of symbols that behaves like a given sequence. The content was coded as patterns, and the representation was coded as P→P because both the stem and the choices use a pictorial representation (figure 5-2).

The two high-performance eighth-grade items in 1995 were also *patterns* items, both of which were also administered to fourth graders. Item L13 is the same as Item L04 above. Item S01a is the first part of a two-part constructed-response item. A sequence of three triangles composed of congruent smaller triangles was shown (figure 5-3). The item asked the student to complete a table by giving the number of triangles composing the second and third figures (a counting task).

The one 1995 high-performance item for twelfth graders was Item D15a. Students are given the graph of time versus speed for a driver who brakes for a cat and then returns by a shorter route. In D15a students read from the graph to find out the maximum speed of the car (figure 5-4).

Table 5-3. *Distribution of TIMSS Absolute High-Performance Algebra Items (p ≥ .75)*

Grade	1995	1999	2003	Total
		Released items		
4	1	—	3	4
8	2	4	2	8
12	1	—	—	1
		Secure items		
4	1	—	3	4
8	2	1	5	8
12	0	—	—	0

Source: Data on performance are from the TIMSS data almanacs. For TIMSS 1995, see timss.bc.edu/timss1995i/data_almanacs_95.html. For TIMSS 1999, see timss.bc.edu/timss1999i/pdf/bialm5mm2.pdf. For TIMSS 2003, see timss.bc.edu/timss2003i/userguide.html.

— Indicates that the grade was not included in the TIMSS assessment that year.

Figure 5-2. *Released Item L04 for Fourth Grade in 1995*

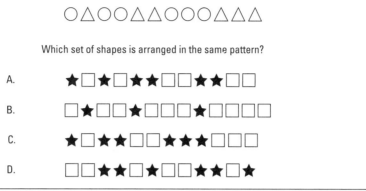

L4. These shapes are arranged in a pattern.

Which set of shapes is arranged in the same pattern?

A.

B.

C.

D.

Source: TIMSS Mathematics Items for the Primary School Years: Released Set for Population 1 (Third and Fourth Grades), p. 32. Available at the TIMSS 1995 home page: timss.bc.edu/timss1995i/TIMSSPDF/AMitems.pdf.

We coded Item D15a's content as dealing with *functions* (because of the relationship between two variables—time and speed—depicted in the graph). It used graphical and verbal representations in the stem and required only a numerical approach (finding the highest point and its measure on the y-axis). We coded the item representation as $VG{\rightarrow}N$.

Figure 5-3. *Triangles for Released Item S01a for Eighth Grade in 1995*

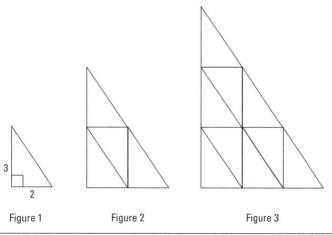

Figure 1   Figure 2   Figure 3

Source: TIMSS Mathematics Items for the Middle School Years: Released Set for Population 2 (Seventh and Eighth Grades), pp. 110–11. Available at the TIMSS 1995 home page: timss.bc.edu/timss1995i/TIMSSPDF/BMItems.pdf.

Figure 5-4. *Graph for Released Item D15a for Twelfth Grade in 1995*

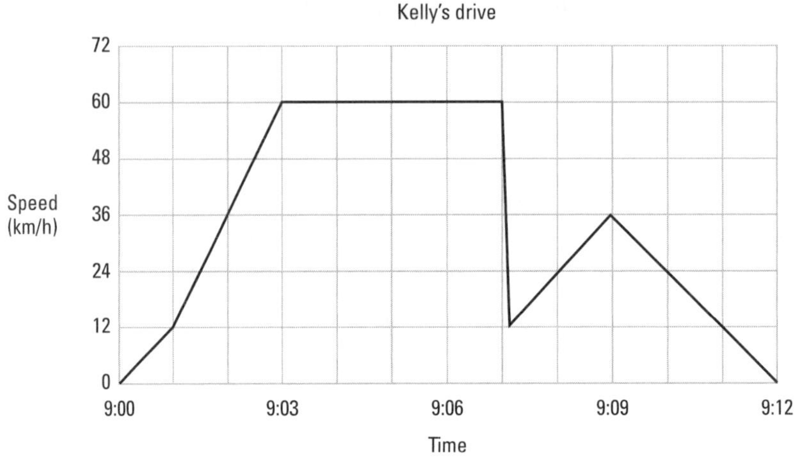

Kelly's drive

Source: TIMSS Mathematics and Science Items for the Final Year of Secondary School: Released Item Set for Population 3, pp. 21–22. Available at the TIMSS 1995 home page: timss.bc.edu/timss1995i/TIMSSPDF/C_items.pdf.

1995 Items: Cognitive Demand. We did not classify as highly demanding the items on which the U.S. students performed well in absolute terms in 1995. The processes for solving these items successfully are quite constrained: reading a simple graph, correctly continuing a pattern, or recognizing it. Items L04, L13, and S01a require mathematical knowledge that is taught in the early grades (for example, pattern recognition in pictorial form or counting figures). Item D15a, in contrast, despite being process constrained, requires knowledge that is acquired only in mathematics classrooms and that is very specific to functional representation. To solve the problem, students must know about graphical interpretation, understand that the representation also explains features of the situation, and observe that the graph also provides a means of monitoring the correctness of the answer.

1999 Items: Content and Representation. In 1999, at least 75 percent of U.S. students—all in the eighth-grade population, the only population assessed— gave a correct response to four released items (as shown in table 5-3). Two of the items, B12 and H12, have the same descriptions as two eighth-grade secure items from 1995.[38] Item B12 in 1999 makes a statement about n (When n is multiplied by 7, and 6 is then added, the result is 41) and asks students to choose the equation that represents the relation. Item H12 provides a situation in words that needs to be modeled with an arithmetic expression ($\square$ represents the number of magazines Lina reads each week), and students are asked to choose the expression that represents the total number of magazines she reads in six weeks. The choices are arithmetic expressions with $\square$ as an operand. Both items use a verbal representation for the statement, but H12 requires recognizing a pre-symbolic representation ($\square$ instead of x). We classified the representations of both, however, as V→S. The other two high-performance items in 1999 correspond to identifying a number that satisfies a proportionality relationship (Item D8: The ratio of 7 to 13 is the same as the ratio of x to 52; what is x?) and identifying a symbolic expression for $n \times n \times n$ (Item P9).

Altogether, the 1999 items on which the U.S. students performed well deal with basic algebraic situations in which students use notions from elementary algebra for the solution. These items tend to involve symbolic representations of verbally presented situations.

1999 Items: Cognitive Demand. In terms of content, the high-performance 1999 items require the use of knowledge that is taught in the middle grades, with the exception perhaps of item H12, which seeks an arithmetic expression that might well have been encountered in the earlier grades. In terms of complexity, Item D8 is probably the most demanding of the group; it involves the use of multiplicative reasoning, which is known to be a difficult conceptual transition for students.[39] The other items ask students to identify a model that matches a given statement, but in Item D8 students might perform any of a number of procedures to get the answer (for example, using the "rule of three," identifying that 52 is obtained by multiplying 13 by 4, and then multiplying 7 by 4, or writing an equation and then cross-multiplying and simplifying). The process in D8, therefore, is less constrained than in the other items.

2003 Items: Content and Representation. In 2003, there were three high-performance items at fourth grade (table 5-3): Items M03_11, M04_04, and M04_05.[40] Item M03_11 asked students to give the three numbers that continue the pattern 100, 1, 99, 2, 98, __, __, __.

Item M04_04 gives students a table with three times at which a movie starts. The student is asked to find the start time for the fourth show assuming that the pattern continues.

Item M04_05 presents a situation verbally ("Ali had 50 apples. He sold some and then had 20 left"). The student is asked to identify the expression (with $\square$ for the unknown number of apples sold) that fits the situation. Note that this item is essentially parallel to Item H12 (Lina's magazines) described above.

The two eighth-grade high-performance items in 2003 (table 5-3) were M01_10 and M03_01. In Item M01_10, the student is given an equation—$12/n = 36/21$—and asked to determine the value of n.

Item M03_01 presents a graph of time versus distance for two hikers. The student is asked to find the time at which the two hikers meet, given that they started from the same place and headed in the same direction.

2003 Items: Cognitive Demand. Like the 1995 items, the 2003 released high-performance items mainly addressed elementary notions of patterns and presymbolization work. Unlike the 1999 items, they did not include algebraic manipulation. As before, the three fourth-grade items were less demanding than the two eighth-grade items. The 2003 items required factual knowledge for which instruction was necessary (such as instruction for dealing with an equation with fractions in which one denominator is unknown, for interpreting a graph, or for operating with time). Students could solve Item M01_10 in at least two ways: by cross-multiplying, or by using a multiplicative argument. Because of these two possibilities, we considered the item to be less process constrained than the other four.

Summary of Absolute High Performance. Table 5-4 provides summary data on the four released high-performance items in 1995, the four in 1999, and the

Table 5-4. *Summary Data on Released Absolute High-Performance Items for Each TIMSS Cycle (p ≥ .75)*

Item	Grade	Content	Representation	Quadrant	U.S. percent correct (SE)
1995					
L04	4	Patterns	$P{\to}P$	III	79 (1.9)
L13	8	Patterns	$P{\to}P$	III	93 (0.8)
S01a	8	Patterns	$P{\to}N$	III	75 (2.2)
D15a	12	Functions	$VG{\to}N$	II	85 (1.0)
1999					
B12	8	Equations	$V{\to}S$	III	78 (1.3)
D8	8	Informal algebra	$V{\to}N$	I	78 (1.2)
H12	8	Informal algebra	$V{\to}S$	III	86 (0.9)
P9	8	Algebraic manipulation	$S{\to}S$	III	85 (1.5)
2003					
M03_11	4	Patterns	$N{\to}N$	III	91 (0.9)
M04_04	4	Patterns	$N{\to}N$	III	83 (1.0)
M04_05	4	Informal algebra	$V{\to}S$	III	84 (1.1)
M01_10	8	Equations	$S{\to}N$	IV	80 (1.1)
M03_01	8	Functions	$VG{\to}N$	IV	80 (1.4)

Source: Data on performance are from the TIMSS data almanacs. For TIMSS 1995, see timss.bc.edu/timss1995i/data_almanacs_95.html. For TIMSS 1999, see timss.bc.edu/timss1999i/pdf/bialm5mm2.pdf. For TIMSS 2003, see timss.bc.edu/timss2003i/userguide.html. Standard errors for 2003 items were provided by Joe Galia, TIMSS & PIRLS International Study Center, Boston College.

For TIMSS 1995 released items, see timss.bc.edu/timss1995i/Items.html. For TIMSS 1999 released items, see timss.bc.edu/timss1995i/Items.html. For TIMSS 2003 released items, see timss.bc.edu/timss2003i/released.html.

five in 2003. For each item, we provide the item number and the grade at which it was administered. We also provide our classification of the item content, the item representation, and the quadrant of the item's cognitive demand. Finally, we give the percent correct for U.S. students given the item (and in parentheses, SE, the standard error).

Across TIMSS cycles and grades, most of the thirteen released high-performance items deal with patterns or informal algebra rather than other content. They most commonly involve numerical representations, especially numerical solutions, and they rarely involve graphical representations—never without accompanying verbal content. Regarding the content–process space classification, the items tend to be those in which the content is relatively lean and the process relatively constrained (Quadrant III). The one item that is both rich and open (demanding), under this classification, is Item D8, an item on proportionality from 1999. The standard errors shown in the table are all relatively low, compared with those for other countries, which indicates that the estimates are relatively precise.

Whether an item is responded to correctly by at least 75 percent of U.S. students tells only part of the story of their high performance in algebra. Even when U.S. students showed high levels of performance, students in other countries often outperformed them. We turn now to the analysis of relative high performance.

Relative High Performance

To analyze relative performance across TIMSS 1995, 1999, and 2003, we chose a set of systems representing comparably developed countries that participated in TIMSS and for which we had complete data. The following countries, all of which are members of the Organization for Economic Cooperation and Development (OECD), met the criteria for participation: For 1995 and 1999, the countries were Australia, Canada, the Czech Republic, Hungary, the Netherlands, New Zealand, and the United States. For 2003, Ontario and Quebec participated as separate systems; they replaced Canada and the Czech Republic. Except for the United States and New Zealand, these countries had all been labeled "A" countries: countries whose performance was statistically significantly higher than the performance of the other countries participating in TIMSS 1995.[41]

We considered the United States to have high performance on an item when it ranked first or second among the seven OECD countries. Across the three cycles, we identified eleven released items (out of eighty-one classified as algebra-related) and nine secure items (out of seventy-six classified as algebra-related) in which the performance of U.S. students ranked first or second. As before, although we looked at performance on the secure items, we do not discuss it here. Table 5-5 shows the distribution of the twenty items across cycles and grades.

Table 5-5. *Distribution of TIMSS Relative High-Performance Algebra Items*

Grade	1995	1999	2003	Total
		Released items		
4	2	—	4	6
8	1	1	3	5
12	0	—	—	0
		Secure items		
4	2	—	2	4
8	0	2	3	5
12	0	—	—	0

Source: See table 5-3.
— Indicates that the grade was not included in the TIMSS assessment that year.

Below we discuss, for each TIMSS cycle, the content, representation, and cognitive demand of the released items for which performance was relatively high.

1995 Items. There were two items, K03 and L09, at fourth grade on which U.S. students ranked first or second in 1995 compared with the other six countries. Item K03 was the following:

Which pair of numbers follows the rule "Multiply the first number by 5 to get the second number"?

    A. 15→3
    B. 6→11
    C. 11→6
    D. 3→15.

The student can either recognize the pair as a fact (and make sure to consider the pair in the right order), check each choice to see whether the second number was obtained when the first was multiplied by 5, or simply note that only one second number is a multiple of 5. Although the content concerns an arithmetic operation, the idea of using one value to obtain another, as in a function machine, is more complex than simply asking for the product of 3 and 5. The notation can also be considered novel. The United States had 65 percent of its fourth graders responding correctly, which was just behind Hungary, which had 68 percent correct. That difference, however, is not statistically significant ($z = .92$, $p > .05$).[42]

Item L09 reads as follows:

Henry is older than Bill, and Bill is older than Peter. Which statement must be true?

    A. Henry is older than Peter.
    B. Henry is younger than Peter.
    C. Henry is the same age as Peter.
    D. We cannot tell who is oldest from the information.

In terms of the cognitive demand, the activity seems constrained (list the names H, B, P; order them according to the given statements $H > B$; $B > P$; and use transitivity to answer the question, $H > P$). The item, however, requires a clear understanding of an important property of natural numbers—that is, transitivity—which is content that students may not be likely to learn without instruction.

In the eighth-grade assessment in 1995, there was only one item, P15, in which the U.S. students' performance was relatively high: second of seven countries. The item reads as follows:

Which of these expressions is equivalent to y^3?

A. $y + y + y$

B. $y \times y \times y$

C. $3y$

D. $y^2 + y$

1999 Items. There was only one relative-high-performance item in 1999, Item P09 (discussed above in the section on absolute performance). Like all the items in that assessment cycle, it was given to eighth graders only.

2003 Items. There were seven released items for which the performance of U.S. students ranked first or second among the seven systems in 2003: four at fourth grade (Items M01_12, M03_11, M04_04, and M04_05); and three at eighth grade (Items M01_12, M04_04, and M09_05).

In the fourth-grade item M01_12, students are given a sentence in words about the number of magazines a child reads in a week, and the item asks students to select the expression that would give the number of magazines read in six weeks. The item uses a verbal representation in stating the task, but requires the use of a semi-symbolic expression ($\square$) and various operations with six for the answer. The process is constrained, and students need to use some arithmetic reasoning to select the appropriate equation. Item M03_11 was the number pattern item given to fourth graders that was discussed in the section on absolute performance, as were Items M04_04, the item containing a table of movie start times, and Item M04_05 (Ali's apples).

Item M01_12, one of the three items given to eighth graders in 2003 on which U.S. students' performance was relatively high, reads as follows:

If $x = -3$, what is the value of $-3x$?

A. -9

B. -6

C. -1

D. 1

E. 9

The process used to solve the item is constrained (direct substitution), but the content—manipulation of negative numbers and substitution of a variable that takes a negative value into an expression that has a negative sign (therefore changing the sign of the expression)—is relatively rich.

Item M04_04 is another 2003 item for which the performance of U.S. eighth graders was relatively high. The item reads as follows:

> The numbers in the sequence 7, 11, 15, 19, 23, . . . increase by four. The numbers in the sequence 1, 10, 19, 28, 37, . . . increase by nine. The number 19 is in both sequences. If the two sequences are continued, what is the next number that is in BOTH the first and the second sequences?

The process used in responding to the item is not very constrained: Students may take the trouble to extend each pattern to find the next number common to both sequences, or they may recall that the greatest common multiple of the two numbers, 36 (because 4 and 9 are relatively prime), gives the next common number after 19 (that is, 36 + 19). They may instead notice that for the two sequences to meet again, their jumps must catch up with each other, which happens after the first sequence has jumped nine times and the second sequence has jumped four times; that is, 36 units after 19. In any case, the mobilization of knowledge and principles is extensive. The task requires some reasoning about sequences even if the patterns are simply extended.

Item M09_05, the third item on which the performance of U.S. eighth graders in 2003 was relatively high, was difficult for students in every country. The item gives the student an algebraic expression, $y = 3x + 2$, and asks which choice expresses x in terms of y. The process is constrained, and the techniques for solving the problem are among the strategies needed in manipulating algebraic expressions.

Summary of Relative High Performance. Table 5-6 provides summary data on the three released relative-high-performance items in 1995, the one in 1999, and the seven in 2003. For each item, we provide the item number and the grade at which it was administered. We also provide our classification of the item content, the item representation, and the quadrant of the item's cognitive demand. Finally, we give the countries ranking first and second, along with the percent correct for each country, the associated standard error, and the results of a statistical test of the difference between percentages.

The items on which the relative performance of U.S. students tended to rank first or second range widely in difficulty, from less than 30 percent correct to more than 90 percent correct, but as a group the items tend to be rather easy. They are mostly elementary items that reflect the transition from arithmetic thinking to algebraic thinking (involving algebraic manipulation, patterns, and informal algebra). They do not involve pictorial or graphical representations. The few items that are relatively rich in content (Quadrants I and II) involve functions or algebraic reasoning, and a couple of them call for numerical representations,

Table 5-6. *Summary Data on Released Relative High-Performance Items for Each TIMSS Cycle*

Item	Grade	Content	Representation	Quadrant	Rank 1	Percent correct (SE)	Rank 2	Percent correct (SE)	p
1995									
K03	4	Functions	$V \to N$	II	Hungary	68 (2.3)	United States	65 (2.3)	n.s.
L09	4	Algebraic reasoning	$V \to V$	II	U.S.	73 (1.7)	Australia	70 (2.0)	n.s.
P15	1	Algebraic manipulation	$S \to S$	III	Czech Republic	85 (2.5)	United States	74 (1.8)	< .01
1999									
P09	8	Algebraic manipulation	$S \to S$	III	U.S.	85 (1.5)	Czech Republic	81 (1.9)	< .05
2003									
M01_12	4	Informal algebra	$V \to S$	III	U.S.	72 (1.2)	Netherlands	72 (2.7)	n.s.
M03_11	4	Patterns	$N \to N$	III	Ontario	93 (1.6)	United States	91 (0.9)	n.s.
M04_04	4	Patterns	$N \to N$	III	U.S.	83 (1.0)	Netherlands	83 (1.9)	n.s.
M04_05	4	Informal algebra	$V \to S$	III	Hungary	84 (1.7)	United States	84 (1.1)	n.s.
M01_12	8	Algebraic manipulation	$S \to N$	III	Hungary	69 (1.9)	United States	66 (1.7)	n.s.
M04_04	8	Algebraic reasoning	$V/N \to N$	I	Hungary	49 (2.3)	United States	45 (1.8)	n.s.
M09_05	8	Algebraic manipulation	$S \to S$	III	U.S.	26 (1.3)	Quebec	24 (2.1)	n.s.

Source: See table 5-4.
n.s. Not significant.

another indicator of the transition from arithmetic to algebra. As we observed in the section on absolute performance, most items are located in Quadrant III of the content–process space, suggesting that they are mostly process constrained and content lean. As an aside, the standard errors for this group of items are somewhat larger than the standard errors for the absolute-high-performance items, indicating a wider range of performance on these items by students in all the systems.

Low-Performance Items

We now consider just those algebra items that no more than 25 percent of the students got correct, quite apart from how students in other countries did on the items.[43] We term this absolute low performance. Then we consider those items on which U.S. students were low performers relative to students in the comparison countries: relative low performance.

Absolute Low Performance (p ≤ .25)

We identified seven released items (out of eighty-one classified as algebra-related) and nine secure items (out of seventy-six classified as algebra-related) that were answered correctly by 25 percent or less of the U.S. students in the sample. Table 5-7 shows the distribution of items across cycles and grades.

1995 Items: Content and Representation. There were three low-performance items in 1995. Item A10 was a mathematics literacy item given to twelfth graders. It gave a pair of axes and an unmarked 16-by-11-square grid, and the student was asked to sketch a graph, with labeled axes and a realistic scale, to represent the relationship between a person's height and age from birth to age 30. The rubric looked for correct labeling of axes, sensible scales, and an adequate growth curve in terms of the two variables.

Table 5-7. *Distribution of TIMSS Low-Performance Algebra Items (p ≤ .25)*

Grade	1995	1999	2003	Total
		Released items		
4	0	—	1	1
8	2	0	3	5
12	1	—	—	1
		Secure items		
4	0	—	2	2
8	0	1	6	7
12	0	—	—	0

Source: See table 5-3.
— Indicates that the grade was not included in the TIMSS assessment that year.

Items S01b and T01a were eighth-grade items. We noted above that U.S. students did very well on S01a (which required the student to count the small triangles composing a larger triangle). In Item S01b, the student is asked to extend the pattern and determine the number of small triangles in the eighth triangle in the sequence, a more challenging mathematical task. The rubric does not indicate that the student needed to come up with a formula for the number of triangles, nor does it suggest that the student use an iterative process to arrive at the solution.

Item T01a asked the student to solve a word problem involving apples in boxes: There are 54 kilograms of apples in two boxes. The second box of apples weighs 12 kilograms more than the first. How many kilograms of apples are in each box? Although the student could have used some symbolic manipulation, the rubric did not suggest that as a possibility.

1995 Items: Cognitive Demand. A major difference between Item A10 and Item D15 (the high-performance twelfth-grade item from 1995 given and discussed above) is that in Item A10 the student had to produce a graph for the situation, a much more difficult task than simply reading the graph.[44] Although the functional relation—that is, that height increases with age—may not be necessarily learned in a mathematics class, the techniques and processes for creating a representation of the problem are. The process for arriving at a particular graph may be limited to recalling milestones or turning points in growth and plotting them on the graph.

Similarly, the difference between Items S01a and S01b resides in the extension of the sequence of triangles in Item S01b to a step in which the student should either formulate a general rule or develop a strategy for systematically counting the triangles at each step. The item can be conceived of as testing the relationship between area and length in similar figures: Altering the lengths of a figure by a factor a affects the area of the resulting similar figure by a factor of a^2. The item can also be conceived of as being about summing the elements in a sequence in which each new figure increases the number of small triangles by an odd number: $a_1 = 1$, $a_2 = 1 + 3$; $a_3 = 1 + 3 + 5$; so the item might be testing whether students could demonstrate that $\Sigma_{k=1,n} (2k - 1) = n^2$ and find the result for $n = 8$. Students might know this relationship already, and then the process would be direct recall. Or in the absence of such knowledge, students might simply calculate each term up to the eighth. The item can also be interpreted as involving a straightforward pattern table in which each second term is the first term squared. Which path a student took might depend on many circumstances. The TIMSS rubric suggests that the important aspect was the final response (64). Because the item requires understanding principles governing these figures, it is a relatively rich, and the process is less constrained than in S01a.

Item T01a requires the identification of the two variables (the number of kilo-grams in each box) and taking into account the constraints on them (altogether there are 54 kilograms, and the difference is 12 kilograms). To respond correctly to the item, a student must know that the two variables are related. The item has the form of a classic problem: The sum of two numbers is a; the difference is b; what are the numbers? Independently of whether students choose to represent the variables with a symbol (such as x and y), they know that the combined weight is 54 (or $x + y = 54$). They also know that one box weighs more than the other ($y = x + 12$). Thus, $2x + 12 = 54$, or $x = 21$, and therefore, $y = 33$. Another approach consists of assuming that the two boxes weighed the same (27 kilo-grams). Because the difference in weights is 12, 6 kilograms need to be taken from one box and put into the other; thus there will be $27 - 6 = 21$ kilograms in one box and $27 + 6 = 33$ kilograms in the other. A student might also use trial and error; that is, looking at how 54 can be decomposed into two integers and taking the pair whose difference is 12 (of course, that process can also be modeled with symbols). These approaches require a level of sophistication that is probably beyond most students. The item also asks students to show work, which only adds to its complexity. The TIMSS rubric credits only the final numerical response as correct and penalizes students when no explanation is provided.

1999 Items. There were no released items in 1999 on which the perfor-mance of U.S. students was low (see table 5-7).

2003 Items: Content and Representation. There were four released items in 2003 to which no more than 25 percent of U.S. students responded correctly (see table 5-7). One item, Item M09_04, was given to fourth graders. It reads as follows:

$$37 \times \square = 703. \text{ What is the value of } 37 \times \square + 6?$$

The other three low-performance items were given to eighth graders. In Item M04_01, the student is given an equation for determining the sum of three consecutive even numbers that add to 84 [$k + (k + 2) + (k + 4) = 84$] and is asked what k represents. The choices are the following: the least of the three numbers, the middle number, the greatest number, or the average.

Item M04_10c is the third part of an item containing three figures: The first is a square with a diagonal forming two triangles; the second is a square whose sides are twice as long and which contains four smaller squares, and therefore eight triangles; and the third is a square whose sides are three times the sides of the first square, so it contains nine squares and eighteen triangles (figure 5-5).

The item is similar to Item S01 in 1995. It has essentially the same problem setup except that it involves twice as many triangles. The first part (a) asks for the number of triangles for the third figure in the sequence (a counting exercise)

Figure 5-5. *Squares for Released Item M04_10c for Eighth Grade in 2003*

The three figures below are divided into small congruent triangles.

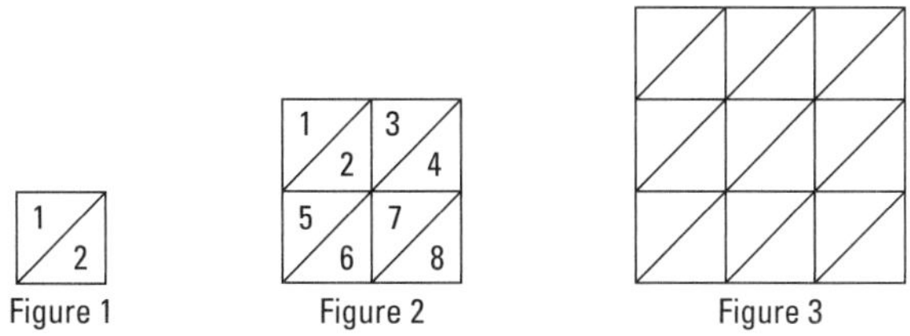

Figure 1   Figure 2   Figure 3

Source: TIMSS 2003 Mathematics Items: Released Set, Eighth Grade, unpaged. Available at the TIMSS 2003 home page: timss.bc.edu/PDF/T03_RELEASED_M8.pdf.

and the fourth figure (either an addition task or, if the general rule has already been obtained, an application of that rule). The second part (b) asks for the number of triangles in the seventh figure ($7^2 \times 2$), and the third part (c) requests an explanation of how to obtain the fiftieth figure "that does not involve drawing it and counting the number of triangles." From that request, we infer that the item is asking for an explicit relationship between the number of the figure in the sequence and the number of small triangles in that figure. It is not asking the student simply to extend the pattern or to find the next three elements in the sequence; instead, it is asking for the function itself. The TIMSS rubric indicates that any of the following should be considered correct: $2n^2$, $2 \times 50 \times 50$, 100×50, $(50 + 50) \times 50$, 2×50^2, or the equivalent expressed in words. Partially correct answers are accepted if the final number is correct.

Item M09_06 states a situation in words involving quantities and costs of two types of fruit and explicitly requests two equations that could be used to find the values of each variable (number and cost).

2003 Items: Cognitive Demand. For all four low-performance items in 2003, the complexity was relatively high. For example, to correctly solve the problem in Item M09_04, students must understand that the equal sign is bidirectional and that an equation relates two objects. They also must know that to maintain the equation, what one does on one side of the equal sign must be done on the other side, which indicates that, in this case, the answer should be $703 + 6$. At a grade—in this case, fourth grade—in which most of the students' arithmetical work deals with operations, the transition to a conception of the equal sign as representing the equivalence of quantities is difficult.[45] Arithmetic and arithmetic sentences are very much involved with *doing*—with processes associated with *operating* on numbers. Moving toward a more static view of the equal sign and seeing expressions linked by that sign as forming an equation constitute a major step toward conceptualizing equations as objects.

Item M04_01 is an unusual item that asks students to consider a symbolic expression and decide on the meaning of the variable. A more common item would have asked the student to find the three numbers, not to give the meaning of the variable. The item draws on students' knowledge of the meaning of algebraic symbols and knowledge of their relation to expressions and equations. That is knowledge for which instruction is necessary, and it requires familiarity with problem representation, but the process is constrained.

The remaining two items, Items M04_10 (sequence of triangles) and M09_06 (number and cost), require more substantial knowledge. Both are constructed-response items and give students the opportunity to select the process by which they will respond. The first item asks for an explanation of how to find a number, and the second asks for two equations that fit a given situation. Both require the coordination of basic knowledge and principles.

Summary of Absolute Low Performance. Table 5-8 provides summary data on the three released low-performance items in 1995 and the four in 2003. For each item, we provide the item number and the grade at which it was administered. We also provide our classification of the item content, the item representation, and the quadrant of the item's cognitive demand. Finally, we give the percent correct and the standard error for U.S. students given the item.

Across TIMSS cycles and grades, the seven released low-performance items tend to deal with equations or functions rather than other content. They mostly involve a variety of representations apart from graphical ones. With respect to the content–process space classification, the items on which U.S. students did

Table 5-8. *Summary Data on Released Absolute Low-Performance Items for Each TIMSS Cycle (p ≤ .25)*

Item	Grade	Content	Representation	Quadrant	U.S. percent correct (SE)
1995					
A10	12	Functions	$V{\to}N$	I	11 (0.8)
S01b	8	Patterns	$P{\to}N$	I	25 (1.6)
T01a	8	Equations	$V{\to}S$	II	25 (1.8)
2003					
M09_04	4	Equations	$S{\to}N$	II	7 (0.7)
M04_01	8	Algebraic manipulation	$VS{\to}V$	II	23 (1.5)
M04_10	8	Functions	$PN{\to}N$ or $PN{\to}S$	I	22 (1.5)
M09_06	8	Equations	$V{\to}S$	I	20 (1.4)

Source: See table 5-4.

not do well in the various TIMSS cycles are all rich in content, and most are open in process. The standard errors suggest that in general the performance on these items is uniformly low, and the variation in performance is small, especially for the most difficult items.

Whether an item is responded to correctly by no more than 25 percent of U.S. students tells only part of the story of their low performance in algebra. When U.S. students showed low levels of performance, students in other countries performed about the same as or did not do as badly as their U.S. peers. We turn now to the analysis of relative low performance.

Relative Low Performance

We considered the United States to have low performance on an item when it ranked last among the seven OECD systems in each cycle. Across the three cycles, we identified seventeen released items (out of eighty-one classified as algebra-related) and eighteen secure items (out of seventy-six classified as algebra-related) in which the performance of U.S. students ranked last. Table 5-9 shows the distribution of the thirty-five items across cycles and grades. Below we discuss, for each TIMSS cycle, the content, representation, and cognitive demand of the released items for which performance was relatively low.

1995 Items. There were no fourth- or twelfth-grade items for which the U.S. students ranked last among the OECD countries in 1995 (see table 5-9), but there were eight released items given to eighth graders on which the U.S. students ranked last: Items I01, J18, L11, L13, P10, S01a, S01b, and T01a.

Item I01 is parallel to the eighth-grade Item M04_01 given in 2003. The text of Item I01 is as follows:

Brad wanted to find three consecutive whole numbers that add up to 81. He wrote the equation $(n - 1) + n + (n + 1) = 81$. What does the n stand for?

Table 5-9. *Distribution of TIMSS Relative Low-Performance Algebra Items*

Grade	1995	1999	2003	Total
		Released items		
4	0	—	0	0
8	8	5	4	17
12	0	—.	—	0
		Secure items		
4	0	—	2	2
8	3	4	7	14
12	2	—	—	2

Source: See table 5-3.
— Indicates that the grade was not included in the TIMSS assessment that year.

There is a slight difference in wording between the two items, but the classification is the same.

Item J18 gives a numerical table representing the relation between x and y, and asks for the missing value for $x = 2$. The student has to find the relationship between the two variables, and it is not straightforward to establish the pattern (for example, the student has to realize that there are missing numbers 3, 5, and 6 for which the relation must be true). The kind of thinking that is needed to solve the problem requires instruction; the process is constrained, but the content is rich.

Item L11 gives a situation of a rubber ball that rebounds half the height each time, starting at 18 meters. The student is asked to find the distance traveled when the ball hits the floor the third time. Some modeling is required, but there is no need to use symbolic representations. The problem can be solved through repeated addition by sketching the heights of the ball, or it can be solved by finding the sum of the numbers in the sequence 18, 9, 9, 4.5, and 4.5. We classified it as an open-process problem with lean content (it uses knowledge acquired in previous grades).

Item L13 is the patterns problem (circles and triangles) that we discussed above in the section on absolute performance (L4 at fourth grade). Item P10 reads as follows:

> If m represents a positive number, which of these is equivalent to $m + m + m + m$?

The remaining three items were discussed in the section on absolute performance. Not much eighth-grade content knowledge is needed for Item S01a; it can be solved using information learned in earlier grades. Item S01b is process open with rich content, and Item T01a is also process open but the content is leaner.

1999 Items. In 1999 there were five released items on which performance of the U.S. students ranked last out of seven countries: Items L12, P11, T01, V04a, and V04b. Like all items in 1999, these were eighth-grade items.

Item L12 resembles Item L11 (bouncing rubber ball) from 1995, discussed above, although it differs from that item in some important ways. Item L12 describes the situation of an elevator that makes a series of starts and stops as it goes from one floor to another. The student is asked to estimate how far the elevator has traveled given that the floors are 3 meters apart. In this item, students have to create a representation of the situation that allows them to interpret correctly the distances that need to be added. It is not a given pattern (as with the rubber ball), so the students have to keep track of where the elevator is. The only operation involved is addition. The process is constrained (there are not many paths to the solution), and the problem demands basic arithmetic operations.

Item P11 is similar to Item P9 from 1995. It reads as follows:

For all numbers k, $k + k + k + k + k$ can be written as:

A. $k + 5$

B. $5k$

C. k^5

D. $5(k + 1)$

Item T01 is a constructed-response item that reads as follows:

A club has 86 members, and there are 14 more girls than boys. How many boys and how many girls are members of the club? Show your work.

This item is similar to Item T01 in the 1995 eighth-grade assessment (dealing with kilograms of apples in two boxes). Students may choose a numerical approach or a symbolic approach, and the TIMSS scoring rubric admits either.

Item V04 gives the student four terms in a sequence of figures involving circles (figure 5-6). The first part asks students to complete a table with the number of circles needed for the fourth figure and then to find the number of circles for the fifth figure. The second part asks for the number of circles in the seventh figure. The parts differ in cognitive demand. Although V04a is essentially a counting task in which the process is constrained, V04b is a more open-process task, as students may choose either to extend the pattern, as in solving V04a, or, if they recognize triangular numbers, by using the formula for the nth triangular number $[n(n + 1)/2]$, or by deriving it (if each figure is visualized as being doubled and arranged in a rectangle, then the number of circles in the figure can be seen as half of $n \times n$ circles plus half of n). These strategies suggest that V04b can be more complex, although nothing precludes the students from doing the repeated addition. In fact, the scoring rubric requires only a numerical answer and does not give credit for the analysis.

Figure 5-6. *Circles for Released Item V04a for Eighth Grade in 1999*

The figures show four sets consisting of circles.

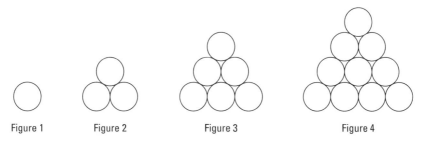

Figure 1   Figure 2   Figure 3   Figure 4

Source: TIMSS 1999 Mathematics Items: Released Set for Eighth Grade, pp. 80–81. Available at the TIMSS 1999 home page: timss.bc.edu/timss1999i/pdf/t99math_items.pdf.

2003 Items. There were four items, all at eighth grade, on which the performance of U.S. students ranked last among the seven systems: Items M01_02, M02_05, M02_06, and M13_05.

Item M01_02 gives a picture portraying a scale that is balanced (figure 5-7). The problem can be solved in at least two ways: numerically or by setting up an equation. It requires the student to use knowledge about balanced scales.

Item M02_05 is an item in which the student is given a picture of matchsticks in a sequence of designs that are increasing in length. The student is given three figures in the sequence and asked for the number of matchsticks in the tenth figure. Because the jump between the third and tenth figures is so great, the students need to derive a relationship to generate the number for the tenth figure. To solve the item correctly, students need to notice that the number of matchsticks increases by three each time. They also have to take into account that the pattern starts with three given matchsticks, so the expression $(3 + 3k)$ gives the number of matches for the kth figure. The process is constrained, and although the task could be solved by repeated addition, the student needs to keep track of the steps.

Item M02_06 requires the student to reason through a situation described in words to determine the equation for the number of books that three students have (figure 5-8). The process is constrained, but it demands that the princi-

Figure 5-7. *Figure for Released Item M01_02 for Eighth Grade in 2003*

The objects on the scale make it balance exactly. One the left pan there is a 1kg weight (mass) and half a brick. On the right pan there is one brick.

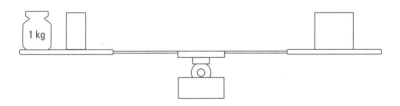

What is the weight (mass) of one brick?

A.	0.5 kg
B.	1 kg
C.	2 kg
D.	3 kg

Source: TIMSS 2003 Mathematics Items: Released Set, Eighth Grade, unpaged. Available at the TIMSS 2003 home page: timss.bc.edu/PDF/T03_RELEASED_M8.pdf.

Figure 5-8. *Figure for Released Item M02_06 for Eighth Grade in 2003*

Graham has twice as many books as Bob. Chan has six more books than Bob.
If Bob has *x* books, which of the following represents the total number of books
the three boys have?

A.	$3x + 6$
B.	$3x + 8$
C.	$4x + 6$
D.	$5x + 6$
E.	$8x + 2$

Source: TIMSS 2003 Mathematics Items: Released Set, Eighth Grade, unpaged. Available at the TIMSS 2003 home page: timss.bc.edu/PDF/T03_RELEASED_M8.pdf.

ples guiding the situation be coordinated to select the appropriate equation. Some manipulation of symbols is required because the expression is given in reduced form.

Item M13_05 is an unusual item. It gives the student a picture of a grid containing a pattern of identical square tiles placed in different orientations so as to make a geometric design (figure 5-9). The student is asked to find the orientation of a tile in a particular square on the grid if the pattern were to be continued. The item requires the coordination of several elements, mostly spatial, to solve the item correctly, but the student also needs to be systematic and to keep track of where the tiles go. There are at least two ways to approach the solution: by labeling the given squares and then repeating the pattern with the labels, or by completing the pattern, either physically or mentally. Some instruction in pattern completion would appear to be necessary if the student is to solve the item correctly.

Summary of Relative Low Performance. Table 5-10 provides summary data on the eight released relative-low-performance items in 1995, the five in 1999, and the four in 2003. For each item, we provide the item number and the grade at which it was administered. We also provide our classification of the item content, the item representation, and the quadrant of the item's cognitive demand. Finally, we give the percent correct and standard error for U.S. students and for students in the next lowest country, together with a test of the significance between the two percentages.

The items on which U.S. performance was relatively low ranged widely in difficulty, from 25 percent correct to 93 percent correct. They tend to be somewhat difficult items in general. Note that all the items on which U.S. performance was relatively low were given to eighth graders, a grade at which much tracking of mathematics classes, both official and unofficial, is in place and at

Figure 5-9. *Figure for Released Item M13_05 for Eighth Grade in 2003*

The tiles can be placed on a grid in four differnt ways. The four ways are shown below with a letter, A, B, C, or D, to identify each one.

A   B   C   D

These letters can be used to describe tiling patterns. For example, the pattern below can be described by the grid of letters shown next to it.

C	A	C	A	C
A	C	A	C	A
C	A	C	A	C

If the pattern on the grid below was continued, what letter would identify the orientation of the tile in the cell labeled X?

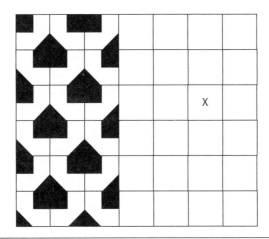

Source: TIMSS 2003 Mathematics Items: Released Set, Eighth Grade, unpaged. Available at the TIMSS 2003 home page: timss.bc.edu/PDF/T03_RELEASED_M8.pdf.

which the relative performance in mathematics of U.S. students has tended to be relatively mediocre in earlier international comparisons. The items require a wide range of skills to be brought to bear on the tasks, and students in the United States perform like students in other countries whose average performance is not strong. The items tend to involve basic skills in arithmetic or algebra.

Table 5-10. Summary Data on Released Relative Low-Performance Items for Each TIMSS Cycle

Item	Grade	Content	Representation	Quadrant	U.S. percent correct (SE)	Next lowest country	Next percent correct (SE)	p
1995								
I01	8	Algebraic manipulation	VS→V	II	32 (1.6)	Canada	34 (2.4)	n.s.
J18	8	Functions	N→N	II	39 (2.1)	Canada	44 (2.3)	n.s.
L11	8	Algebraic reasoning	V→N	IV	27 (1.9)	Netherlands	38 (3.8)	n.s.
L13	8	Patterns	P→P	III	93 (0.8)	Hungary	93 (1.3)	n.s.
F10	8	Algebraic manipulation	S→S	III	46 (2.5)	Netherlands	51 (4.5)	n.s.
S01a	8	Patterns	P→N	III	75 (2.2)	Czech Republic	75 (2.4)	n.s.
S01b	8	Patterns	P→N	I	25 (1.6)	New Zealand	31 (2.5)	<.05
T01a	8	Equations	V→N	II	25 (1.8)	New Zealand	27 (2.0)	n.s.
1999								
L12	8	Algebraic reasoning	V→N	III	55 (1.7)	Netherlands	58 (4.7)	n.s.
P11	8	Algebraic manipulation	S→S	II	46 (1.6)	New Zealand	54 (2.9)	<.01
T01	8	Equations	V→N	I	35 (1.1)	New Zealand	38 (2.3)	n.s.
V04a	8	Patterns	P→N	III	73 (1.4)	Hungary	77 (1.7)	<.05
V04b	8	Patterns	P→N	III	64 (1.4)	New Zealand	64 (2.0)	n.s.
2003								
M01_02	8	Equations	P→N	II	74 (1.3)	New Zealand	75 (2.4)	n.s.
M02_05	8	Patterns	P→S	II	56 (1.8)	Netherlands	59 (1.8)	n.s.
M02_06	8	Equations	P→P	I	26 (1.4)	New Zealand	28 (2.2)	n.s.
M13_05	8	Patterns	V→S	II	48 (2.4)	Australia	51 (3.0)	n.s.

Source: See table 5-4.
n.s. Not significant.

Items that deal with patterns are prominently represented in table 5-10. In every case, the pattern may be continued for a few steps, but a solution to the problem requires some generalization beyond what can be reasonably accomplished by counting or repeated addition. This challenge seems to be a weakness for U.S. students. Interestingly, however, in most cases, students in the other countries tended to perform at about the same level.

Items in Quadrant II are also common in table 5-10. Such items may have a constrained process, but they require relatively high levels of content for their solution, and one might have predicted low performance by U.S. eighth graders. Items in Quadrant III (lean in content and process constrained) are also prominently represented, which is an interesting result. Students in most of the countries tended to do relatively well on those items (except for Items P10 and L12). Items in Quadrants I and IV, which require more open processes, tended to be more difficult too (performance between 25 and 50 percent). In only one case, however, is the performance of the U.S. students on an item in Quadrant I or IV significantly below that of the next lower country, and consequently the other countries in the group. The relatively few significant differences in the table, along with the variety of "next lowest" countries, suggest that U.S. students' performance is not markedly below that of other countries. Notice also that the standard errors tend to be much larger for other countries than for the United States, suggesting a less uniform performance in those countries. The observed differences across countries appear to be attributable to different curricular intentions and implementations.

Content and Process of High- and Low-Performance Items

Figure 5-10 shows the location in the content–process space of those items on which U.S. students' performance was high or low, either absolutely or relative to that of the other OECD countries. All of the items on which performance was low in an absolute sense were rich in content (Quadrant I or II), as were most of those on which performance was relatively low. In contrast, items on which performance was high were much less likely to be rich in content. The process in those items on which performance was relatively low tended to be constrained (Quadrant II or III) rather than open, and the process in the high-performance items, whether absolutely or relatively, was also more likely to be constrained than open.

Figure 5-11 shows items grouped by the representation used in the item stem. The figure shows that more items tended to be presented verbally or pictorially than symbolically or numerically, and that U.S. students tended not to perform well on the pictorial items. It was also the case that more items tended to require a numerical or symbolic response than a pictorial or verbal response

Figure 5-10. *Percent of Items in Each Quadrant by Level of Performance*

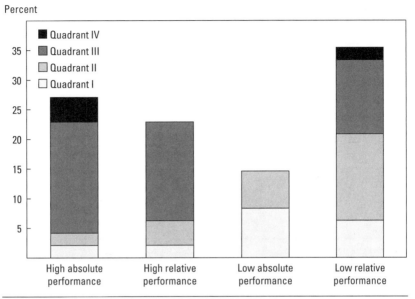

Source: Data are from tables 5-4, 5-6, 5-8, and 5-10.

Figure 5-11. *Number of High- or Low-Performance Items by Representation Used in the Stem*

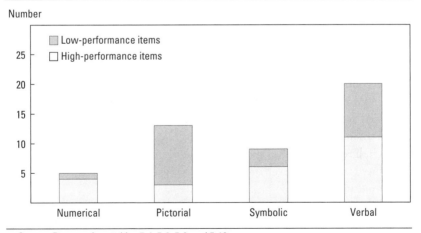

Source: Data are from tables 5-4, 5-6, 5-8, and 5-10.

Figure 5-12. *Number of High- or Low-Performance Items by Representation of Required Response*

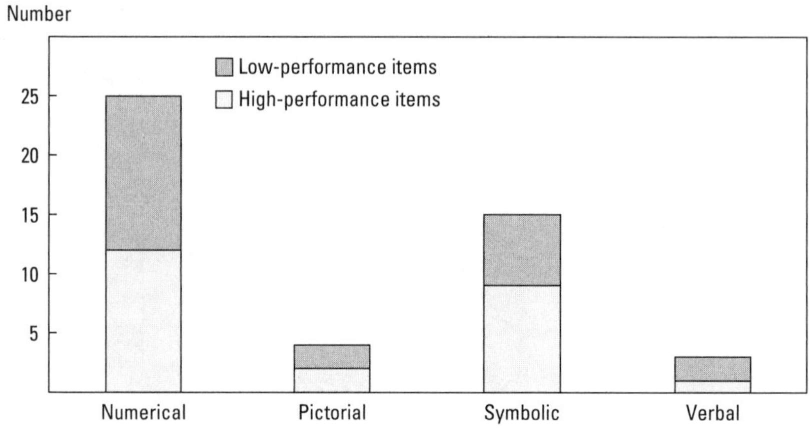

Source: Data are from tables 5-4, 5-6, 5-8, and 5-10.

(figure 5-12). But differences in U.S. students' performance across the type of response required were not strong.

Summary Observations about U.S. Algebra Performance

The level of performance of U.S. fourth graders is not bad on items that ask for rather straightforward calculations or reasoning. They do relatively well in interpreting a rule, engaging in reasoning, translating from words to symbols, and extending numerical patterns. Part of this proficiency may come from the early introduction of ideas of patterns in the U.S. curriculum. Patterns (but not functions per se) are introduced quite early relative to the high-achieving TIMSS countries: according to Valverde and Schmidt, the topic of *patterns, relations, and functions* is introduced in the United States somewhere during the first through third grades, while in the highest-achieving countries, the topic is not introduced until grade seven or eight, and then "most likely" at a different level of complexity.[46]

U.S. twelfth graders do rather well in interpreting the graph of a function, but relative to twelfth graders in other countries, their performance is weak.

U.S. eighth graders appear to do better than students in other countries on items requiring algebraic manipulation when the items tend to be lean in content and process constrained (Quadrant III). When the content of the items is richer, they tend to perform relatively worse than students in the other countries analyzed. U.S. eighth graders also have trouble with equations items, both in relative and in absolute terms, when the items have a high cognitive demand, and especially when they have rich content (Quadrants I and II). This pattern is perhaps another indication that the U.S. students are relatively less exposed to algebraic content than students in other TIMSS countries.

On items classified as informal algebra, U.S. students do well at both fourth and eighth grade. Most of the informal algebra items in the analysis, however, were classified as low demand (lean in content and constrained in process). On items we classified as dealing with patterns, the U.S. students' performance was mixed. On about half of the pattern items, U.S. students performed high in both absolute and relative terms, with most such items classified in Quadrant III (lean content and constrained process) and about a quarter classified as content rich. This result may appear somewhat surprising until one considers that on all but one of the Quadrant III pattern items, the U.S. students performed at a high absolute level but low *relative* to students in other countries. These items tended to require an extension or a description of a pattern, and not merely that the pattern be recognized— a much easier task for U.S. students.

Only one of the items we classified as *Equations* or *Functions* was in Quadrant III, and on only a third of the items was the performance of U.S. students high in both absolute and relative terms. Most of the items were in Quadrants I and II, suggesting that these items had rich content; again the low performance, in both absolute and relative terms, may indicate the lack of exposure of the U.S. students to this content.

The most and greatest differences between performance in the United States and in other countries come at eighth grade. U.S. eighth graders demonstrate relatively good understanding of the notation for exponents, ability to interpret simple algebraic expressions, reasoning about sequences, and solving an equation for one variable in terms of another. In contrast, their performance is relatively weak in interpreting symbols in an equation, completing a table showing a relation between variables, finding the sum of series expressed by verbal rules, identifying a pattern, manipulating a simple algebraic expression, extending sequences of geometric figures to find the pattern, solving word problems involving relations between quantities, translating from words into algebraic symbols, and completing a geometric pattern. As a group, U.S. eighth graders do better in algebra than might be expected given that most of them have limited exposure to it.

To some degree, our results extend and complement those of a study comparing the performance of U.S. students with that of an average taken across twelve countries participating in TIMSS 2003 and PISA 2003.[47] When items from TIMSS were classified as low rigor versus high rigor and from both TIMSS and PISA as low difficulty versus high difficulty, U.S. performance was below the twelve-country average at both levels of rigor and at both levels of difficulty. Our results indicate, however, that although U.S. students' performance may be low in general, it is not uniformly low.

Conclusion

Given that U.S. students' performance in algebra is roughly at their average across mathematics topics, we chose to examine the extremes of the distribution of item difficulty to see what U.S. students manage quite well, what they need help with, and how their performance compares with that of students in other systems. U.S. eighth graders are reasonably competent with problems that involve algebraic manipulation and the transition from arithmetic to algebra (which might be considered good news, given that most U.S. students do not study much algebra before ninth grade). But U.S. students do not do well on items that involve the extension of a pattern if the item requires that they explicitly produce, describe, or represent a relationship rather than simply find the next few terms in a sequence.

U.S. eighth graders have teachers who claim—far more than teachers in other countries—that in more than half of their lessons, they relate the mathematics to daily life.[48] Their students, however, do relatively poorly in setting up an equation to model a real situation. Compared with eighth-grade teachers in some other countries, U.S. teachers tend not to use many high-complexity problems in their lessons, which may help account for some of the difficulties that U.S. students have.[49] Beyond tracking, another source of poor performance in algebra may be that U.S. eighth-grade teachers spend considerable time reviewing topics already taught; almost 30 percent of their lessons are devoted entirely to review.[50]

In short, if they are to improve their performance in algebra, U.S. students appear to need many more opportunities to engage in functional thinking with complex problems and in particular, in functional thinking as it relates to realistic situations.[51] They will live not in a postindustrial society but rather in a knowledge society that will demand a different set of skills than mathematics courses have traditionally supplied.[52] These students need to be able to use the algebra that they study to solve problems they will face professionally and personally.

Algebra is of limited use if it is understood as generalized arithmetic only. If students are to use algebra fluently, they need to be proficient in functional thinking—even if an assessment such as TIMSS does not tap much of that thinking.[53] The algebra they study in school should enable them not simply to manipulate expressions and solve equations but also to formulate problems using algebraic notation, fit functions to data, manipulate those functions to understand phenomena, visualize functional relations, and interpret properties of functions. The United States is not the only country in which eighth-grade teachers could be giving greater attention to functions, but it is one in which too many people have assumed for too long that most students cannot learn, use, or value algebra.

Notes

1. For examples of such advocacy, see Daniel Chazan, *Beyond Formulas in Mathematics and Teaching: Dynamics of the High School Algebra Classroom* (New York: Teachers College Press, 2000), and M. Kathleen Heid, "A Technology-Intensive Functional Approach to the Emergence of Algebraic Thinking," in *Approaches to Algebra: Perspectives for Research and Teaching,* edited by Nadine Bednarz, Carolyn Kieran, and Lesley Lee (Dordrecht: Kluwer Academic Publishers, 1996), pp. 239–55.
2. Robert E. Reys, "Curricular Controversy in the Math Wars: A Battle without Winners," *Phi Delta Kappan* 83 (November 2001): 255.
3. William H. Schmidt, "Are There Surprises in the TIMSS Twelfth Grade Results?" Press Statement by William H. Schmidt, U.S. TIMSS National Research Coordinator, Michigan State University, February 1998 (ustimss.msu.edu/12gradepr.htm).
4. See, for example, Organization for Economic Cooperation and Development, *Learning for Tomorrow's World: First Results from PISA 2003* (Paris: OECD, 2004).
5. Ina V. S. Mullis and others, *TIMSS Assessment Frameworks and Specifications,* 2nd ed. (Boston College, 2003), p. 4.
6. Ibid., p. 3.
7. David P. Baker and Gerald K. LeTendre, *National Differences, Global Similarities: World Culture and the Future of Schooling* (Stanford University Press, 2005), p. 162.
8. The U.S. undergraduate curriculum is a four-year curriculum: freshman, sophomore, junior, and senior. For a fuller account of the formation of the U.S. secondary school mathematics curriculum, see George M. A. Stanic and Jeremy Kilpatrick, "Mathematics Curriculum Reform in the United States: A Historical Perspective," *International Journal of Educational Research* 5 (1992): 407–17.
9. Florian Cajori, "The Teaching and History of Mathematics in the United States," Bureau of Education Circular of Information 3 (U.S. Government Printing Office, 1890), p. 57.
10. Jeremy Kilpatrick, "Confronting Reform," *American Mathematical Monthly* 104 (December 1997): 955–62.
11. Lynn Arthur Steen, "Algebra for All in Eighth Grade: What's the Rush?" *Middle Matters* 8 (Fall 1999): 1, 6–7.
12. Edward A. Silver, "Improving Mathematics in Middle School: Lessons from TIMSS and Related Research" (U.S. Department of Education, 1998), p. 3; see also William H. Schmidt, Hsing Chi Wang, and Curtis C. McKnight, "Curriculum Coherence: An Examination of U.S. Mathematics and Science Content Standards from an International Perspective," *Journal of Curriculum Studies* 37 (2005): 525–59.
13. U.S. Department of Education, "Mathematics Equals Opportunity: White Paper Prepared for U.S. Secretary of Education Richard W. Riley" (October 20, 1997), p. 17 (www.ed.gov/pubs/math/mathemat.pdf).
14. National Assessment Governing Board (NAEP), "Mathematics Framework for the 2005 National Assessment of Educational Progress" (U.S. Department of Education, September 2004), p. 10.
15. John Wirt and others, "The Condition of Education 2004," NCES 2004-077 (U.S. Department of Education, National Center for Education Statistics, 2004), pp. 70, 85.

16. See, for example, Jon R. Star, Beth A. Herbel-Eisenmann, and John P. Smith III, "Algebraic Concepts: What's Really New in New Curricula?" *Mathematics Teaching in the Middle School* 5, no. 7 (2000): 446–51.

17. Regardless of how the approach to algebra taken in the U.S. curriculum is characterized, one can argue that it has become increasingly demanding, like that of the school mathematics curriculum in general over the past century; see, for example, Clancy Blair, and others, "Rising Mean IQ: Cognitive Demand of Mathematics Education for Young Children, Population Exposure to Formal Schooling, and the Neurobiology of the Prefrontal Cortex," *Intelligence* 33 (2005): 93–106.

18. Ina V. S. Mullis, Michael O. Martin, and Dana Diaconu, "Item Analysis and Review," in *TIMSS 2003 Technical Report,* edited by Michael O. Martin, Ina V. S. Mullis, and Steven J. Chrostowski (Boston College, 2004).

19. Data and analyses for TIMSS 1995 can be found in reports by Mary M. Lindquist and Jeremy Kilpatrick for the National Council of Teachers of Mathematics that were posted anonymously on the NCTM website immediately after each release of TIMSS data (for example, www.nctm.org/news/releases/1996/timss_eighth_grade.htm). See also National Center for Education Statistics, "Pursuing Excellence: A Study of U.S. Eighth-Grade Mathematics and Science Teaching, Learning, Curriculum, and Achievement in International Context," NCES 97-198 (U.S. Department of Education, November 1996); "Pursuing Excellence: A Study of U.S. Fourth-Grade Mathematics and Science Achievement in International Context," NCES 97-255 (U.S. Department of Education, June 1997); and "Pursuing Excellence: A Study of U.S. Twelfth-Grade Mathematics and Science Achievement in International Context," NCES 98-049 (U.S. Department of Education, August 1998). For TIMSS 1999, see Patrick Gonzales and others, "Highlights from the Third International Mathematics and Science Study–Repeat (TIMSS–R)," NCES 2001-027 (U.S. Department of Education, National Center for Education Statistics, December 2000).

20. Michael O. Martin and Ina V. S. Mullis, "Overview of TIMSS 2003," in *TIMSS 2003 Technical Report,* edited by Martin, Mullis, and Chrostowski, p. 18.

21. Patrick Gonzales and others, "Highlights from the Trends in International Mathematics and Science Study: TIMSS 2003," NCES 2005-005 (U.S. Department of Education, National Center for Education Statistics, December 2004), p. 4.

22. Rodger W. Bybee and others, "TIMSS 2003: An Introduction," *The Natural Selection: The Journal of BSCS* (Winter 2005): 9.

23. Patrick Gonzales and others, "Highlights from TIMSS 2003," p. 5.

24. Bybee and others, "TIMSS 2003: An Introduction," p. 9.

25. Gonzales and others, "Highlights from TIMSS 2003," p. 10.

26. Jeremy Kilpatrick, Jane Swafford, and Bradford Findell, eds., *Adding It Up: Helping Children Learn Mathematics* (National Academy Press, 2001), p. 280.

27. Alan Ginsburg and others. "Reassessing U.S. International Mathematics Performance: New Findings from the 2003 TIMSS and PISA" (American Institutes for Research, November 2005).

28. Jeremy Kilpatrick and Judith Lynn Gieger, "The Performance of Students Taking Advanced Mathematics Courses," in *Results from the Seventh Mathematics Assessment of the National Assessment of Educational Progress,* edited by Edward A. Silver and Patricia A. Kenney (Reston, Va.: National Council of Teachers of Mathematics, 2000), pp. 377–409.

29. See Nicolas Balacheff and Claire Margolinas, "Modèle de connaissances pour le calcul de situations didactiques" [Model of Conceptions for Defining Didactic Situations], in *Balises pour la didactique des mathématiques,* edited by Alain Mercier and Claire Margolinas (Grenoble: La Pensée Sauvage, 2005); and Shlomo Vinner, "Concept Definition, Concept Image and the Notion of Function," *International Journal of Mathematics Education in Science and Technology* (1983): 293–305.

30. These categories are used by NAEP ("Mathematics Framework for 2005," p. 34) to describe forms of algebraic representation; we have combined somewhat different forms than those NAEP uses at grades four, eight, and twelve.

31. Gail P. Baxter and Robert Glaser, "Investigating the Cognitive Complexity of Science Assessments," *Educational Measurement* 17, no. 3 (1998): 37–45.

32. Ibid., p. 38.

33. Ibid.

34. See J. Richard Landis and Gary G. Koch, "The Measurement of Observer Agreement for Categorical Data," *Biometrics* 33 (1977): 159–74.

35. See Martin, Mullis, and Chrostowski, *TIMSS 2003 Technical Report,* for information on TIMSS data almanacs.

36. "The released items are pretty much representative of the total item pool" (personal communication from Michael O. Martin, Boston College, TIMSS & PIRLS International Study Center, January 23, 2007).

37. The "least difficult quartile" is an item category used by Ginsburg and others, "Reassessing U.S. Mathematics Performance."

38. The items have the same IDs as the 1995 items and are likely to be identical to their secure counterparts in view of the TIMSS policy of writing new items and releasing some secure items for each successive cycle.

39. Gerard Vergnaud, "Multiplicative Conceptual Fields: What and Why?" in *The Development of Multiplicative Reasoning,* edited by Guershon Harel and Jere Confrey (SUNY Press, 1994), pp. 41–59.

40. In 2003, there was a strong position effect for some TIMSS items because they appeared in different positions in the booklets. "Some students in all countries did not reach all the items in the third block position, which was the end of the first half of each booklet before the break. The same effect was evident for the sixth block position, which was the last block in the booklets" (Mullis, Martin, and Diaconu, "Item Analysis and Review," p. 248). Items in the third and sixth block positions were treated in TIMSS scale analyses as if they were unique, but we were advised not to include those items in our study (personal communication from Joe Galia, Boston College, TIMSS & PIRLS International Study Center, October 11, 2006).

41. Gilbert A. Valverde and William H. Schmidt, "Greater Expectations: Learning from Other Nations in the Quest for 'World-Class Standards' in U.S. School Mathematics and Science," *Journal of Curriculum Studies* 32, no. 5 (2000): 615–87.

42. We used the following formula for the difference between two proportions taken from independent populations: $z = (p_1 - p_2)/\text{sqrt} (\text{SE}_1{}^2 + \text{SE}_2{}^2)$, where p_i is the proportion of students who responded correctly to the item in country i, and SE_i is the standard error for that item and country. $z \sim N (0, 1)$. We conducted paired comparisons only (to test whether the U.S. students were in first or second place). We did not engage in multiple comparisons (for example, to test whether the U.S. students scored differently from a country in third place), which would have made the exposition problematic. Put simply, all we wanted to know was whether the U.S. students ranked first or second (even when that ranking might have been shared with another country).

43. The "most difficult quartile" is an item category used by Ginsburg and others, "Reassessing U.S. Mathematics Performance."

44. Kilpatrick, Swafford, and Findell, *Adding It Up,* p. 290.

45. Carolyn Kieran, "Concepts Associated with the Equality Symbol," *Educational Studies in Mathematics* 12 (1981): 317–26.

46. Valverde and Schmidt, "Greater Expectations," p. 678.

47. Ginsburg and others, "Reassessing U.S. Mathematics Performance."

48. Ibid., p. 23.

49. James Hiebert and others, "Highlights from the TIMSS 1999 Video Study of Eighth-Grade Mathematics Teaching," NCES 2003-011 (U.S. Department of Education, National Center for Education Statistics, 2003), p. 6; Iris Weiss, Joan D. Pasley, P. Sean Smith, Eric R. Banilower, and Daniel J. Heck, *Looking inside the Classroom: A Study of K-12 Mathematics and Science Education in the United States* (Chapel Hill, N.C.: Horizon Research, 2003), p. 40.

50. Hiebert and others, "Highlights from the TIMSS Video Study," p. 5.
51. The term *functional thinking* (*funktionales Denken*) was introduced by the mathematician Felix Klein at a 1905 conference in which he argued for reorganizing the secondary mathematics curriculum around the concept of function. See Martin A. Nordgaard, "Introductory Calculus as a High School Subject," in *Selected Topics in the Teaching of Mathematics,* 3rd Yearbook of the National Council of Teachers of Mathematics, edited by John R. Clark and William D. Reeve (Teachers College, Columbia University, Bureau of Publications, 1928), pp. 65–101. Functional thinking meant that students should learn to think in terms of variables and the functions that describe how one variable depends upon others.
52. Andy Hargreaves, *Teaching in the Knowledge Society: Education in the Age of Insecurity* (Teachers College Press, 2003).
53. *Patterns, relations, and functions* was among the top five topics included in eighth-grade standards documents and textbooks from thirty-six countries participating in TIMSS 1995, but it did not appear among the top five topics that teachers reported teaching or to which the most teaching time was devoted. See Leland S. Cogan and William H. Schmidt, " 'Culture Shock'—Eighth-Grade Mathematics from an International Perspective," *Educational Research and Evaluation* 8 (2002): 37.

6

What Can TIMSS Surveys Tell Us about Mathematics Reforms in the United States during the 1990s?

LAURA S. HAMILTON AND JOSÉ FELIPE MARTÍNEZ

Throughout the 1990s, a number of mathematics reforms were introduced in kindergarten through grade twelve throughout the United States. Many of these reforms were influenced by the National Council of Teachers of Mathematics (NCTM), which published national standards for mathematics curriculum and instruction, and by the National Science Foundation, which funded the development of curriculum materials aligned to the NCTM Standards.[1] These reforms and many of the new curricula emphasized the importance of problem solving and inquiry and were designed to promote increased use of instructional activities that reformers believed would foster students' thinking skills. These activities included, for example, cooperative groups, writing about mathematics, and use of open-ended assessment strategies. Reform advocates also pointed to the value of studying topics in depth rather than covering a large number of topics in a superficial way, and they encouraged certain uses of technology, including calculators, when the technology served to promote deeper knowledge and understanding. Although NCTM did not endorse instructional approaches that ignored the development of basic skills and factual knowledge, and has published several subsequent documents that clarify this position, critics of the reforms expressed concerns that certain types of knowledge and skills were often deemphasized.[2] Looking back on this decade of reform, it is important to ask whether and how these reforms might have influenced teachers' practices and student achievement in mathematics.

In this paper, we review work that has examined relationships between students' mathematics achievement and teachers' use of so-called "reform-oriented" instructional practices. We then present new analyses using data from the Trends in International Mathematics and Science Study (TIMSS) to examine teachers' use of these practices in the United States and in a small number of other countries, and their relationships with student achievement in mathematics. We also discuss ways in which existing research and data sources fall short of answering the most important questions about reform-oriented instruction.

Distinguishing Reform from Traditional Practices: Understanding the Math Wars

For the past several decades, discussions about mathematics curricula have often been characterized by conflict between two camps—advocates of what is often called "traditional" instruction on the one side, and reformers who espouse student-centered or inquiry-based approaches on the other. Proponents of more traditional approaches often argue that students learn mathematics most effectively when given extensive opportunities to develop and practice skills and when provided with direct instruction from a teacher. They point to low student achievement, particularly on tests that measure computational skills, and note the risks this poor performance poses for these students' future mathematical performance and for the nation's economy.[3] The reform-oriented approach, by contrast, places greater emphasis on having students construct their own knowledge through investigations and other student-led activities.[4] Reform-minded educators and scholars often cite low test scores as evidence that the traditional approach that has been prevalent in U.S. schools has not worked.[5] Some members of each camp have assigned labels to the other side, with traditionalists being labeled as in favor of "drill and kill" approaches, and reform advocates being accused of promoting "fuzzy math." To illustrate, Chester Finn, who has argued for the importance of teaching basic skills, was quoted in a *New York Times* article, referring to "the constructivist approach some educators prefer, in which children learn what they want to learn when they're ready to learn it."[6] Finn's view represents a distortion of the kind of constructivist approach that has been thoughtfully described by scholars such as Battista, but it is not an uncommon one.[7]

There are signs of a partial cease-fire in the math wars. NCTM has published a number of documents that clarify its own positions, most recently the *Curriculum Focal Points for Prekindergarten through Grade 8 Mathematics,* which includes references to the importance of computation and other basic skills.[8] Although NCTM officials argue that this has been their position all along, many observers of the math wars perceive this and related documents as indicators of a growing acceptance of more traditional approaches on the part of NCTM.[9] The National Mathematics Advisory Panel, created in April 2006 by President George W. Bush to provide advice on the effectiveness of various approaches to teaching mathematics, includes individuals who have been associated with both sides of the debate. Early indications of the panel's work suggest a greater degree of consensus than has been obtained in the past.

Despite the often rancorous environment that characterizes the debate, the two approaches do not necessarily operate in opposition to one another. Moreover, the notion that "reform" and "traditional" approaches can be easily distinguished from one another is simplistic. Most of those who espouse the traditional viewpoint do not believe that students should engage in mindless drills or that higher-order reasoning is unimportant, and the idea that reform-oriented instruction shuns computation in favor of purely "fuzzy" activities that put the students in charge does not consider the ways in which many NCTM-aligned curricula incorporate a variety of activities including computation practice and teacher-led instruction. Interest in reform-oriented instruction is often grounded in a belief that while students are likely to benefit from traditional approaches, these approaches alone are insufficient: Students also need to be exposed to instruction that promotes conceptual understanding and the ability to solve complex, ill-structured problems, that encourages them to communicate mathematically and to ask questions, and that helps them make connections between mathematics and other disciplines.[10] The relevance of traditional approaches and goals to effective reform instruction is described by Battista:

> It is clearly insufficient to involve students only in sense making, reasoning, and the construction of mathematical knowledge. Sound curricula must have clear long-range goals for assuring that students become *fluent* in utilizing those mathematical concepts, ways of reasoning, and procedures that our culture has found most useful. . . . They should possess knowledge that supports mathematical reasoning. For example, students should know the "basic number facts" because such knowledge is essential for mental computation, estimation, performance of computational procedures, and problem solving.[11]

Investigations about the effectiveness of reform-oriented instruction, therefore, need to recognize that the use of more traditional practices is not antithetical to reform and is likely to occur concurrently with reform-oriented practices in many classrooms. In fact, efforts to measure teachers' use of both approaches typically reveal either null or positive correlations between reform and traditional approaches.[12] Moreover, the field has not reached consensus on how to define "reform-oriented" or "traditional" instruction, so while there is extensive overlap across studies in how these constructs are conceptualized and measured, they do not always mean exactly the same thing.

This discussion makes clear that simple comparisons of reform versus traditional practices are unlikely to shed much light on questions about the effectiveness of various teaching strategies, given the complex ways in which the two approaches interact and the lack of a universally accepted definition of either approach. Nonetheless, it is worth examining the existing data and literature to understand how frequently teachers use various instructional approaches and how these approaches are associated with student achievement. Before we turn to a new set of analyses using TIMSS data we briefly review a number of studies that have addressed this topic in recent years.

Existing Research on Reform Practices

The research literature suggests that only a small minority of U.S. teachers embody the principles of reform-oriented curriculum and instruction in their lessons. In one recent study, Jacobs and colleagues used the 1995 and 1999 TIMSS video studies to examine various dimensions of reform practice among middle-school teachers and found that the typical teacher used few reform-oriented practices.[13] Moreover, that study's comparison of lessons in 1995 and 1999 suggested that changes in fidelity to reform practices during that period were relatively few. The changes that were observed went in both directions, with teachers in 1999 slightly more likely to adopt some aspects of the reforms and less likely to adopt others. Other work lends support to the conclusion that reform-oriented instruction is relatively rare compared with more traditional approaches and indicates that even when teachers are given reform-oriented curriculum materials, they often fail to implement the curriculum in ways consistent with developers' intentions.[14] The relative infrequency of reform-oriented instruction is not unique to the United States: one recent international comparison using 1999 eighth-grade TIMSS survey data showed that the rate of use among U.S. teachers is similar to the international average.[15]

One reason for the relatively infrequent use of reform practices might be the lack of a strong research base for concluding that these practices promote improved achievement. Research in cognitive and developmental psychology provides some evidence that student learning in mathematics is associated with exposure to aspects of reform-oriented instruction, such as the opportunity for students to construct their own ideas about mathematics.[16] Most of this work was done in a relatively small number of classrooms, which made it possible to collect rich data on instruction and achievement but limited the generalizability of results.

Several studies examining how teachers' use of reform practices is related to student achievement in large numbers of schools and classrooms have also been conducted.[17] Together, the findings can be described as suggesting, at best, a small positive relationship between some aspects of reform-oriented instruction and achievement, but the body of work is far from conclusive, and many of the studies also point to positive relationships with traditional practices as well. Most of this work relies on surveys that ask teachers to report the frequency of various instructional activities such as use of cooperative groups and administration of open-ended assessments. Undoubtedly such surveys fail to capture some of what distinguishes true reform-oriented instruction from other approaches, a problem to which we return in the final section of this paper. In particular, content is often ignored in these studies in favor of a focus on pedagogy. Nonetheless, these survey-based studies are the source of much of what we currently know about how instructional practices are related to student achievement. Moreover, several studies suggest that questionnaires can provide reasonably accurate information about practices and that acceptable levels of agreement between questionnaire responses and observations are often obtained.[18]

Limitations in the way student achievement is measured also affect conclusions made from research on instructional practices. Researchers have called attention to the possibility that different outcomes or constructs may exhibit different relationships with a common set of instructional practices, even within the same content area. Even with a test that functions in a unidimensional manner, conclusions about the relationship between mathematics achievement and instruction have been shown to depend on the specific choice of items included or the weights assigned to each subtopic in the test.[19] Estimates of relationships between reform-oriented instruction and achievement are likely to depend in part on the degree to which the outcome measure is sensitive to the specific instruction offered.[20]

Using TIMSS Surveys to Examine Reform-Oriented Instruction

As discussed earlier, TIMSS survey data have previously been used to investigate teachers' use of reform-oriented instruction in mathematics. In addition, the TIMSS surveys have been adapted for use in other studies of instructional practices.[21] In this paper we present the results of several analyses using the 2003, 1999, and 1995 eighth-grade TIMSS data, with a focus on understanding the extent to which teachers in the United States and elsewhere have adopted reform-oriented instructional methods, and whether use of these methods in the classroom is associated with differences in student achievement.

We address four questions using data from the United States and the other participating countries:

—How has teachers' reported use of reform-oriented instruction changed over time, including their use of reform-oriented pedagogical practices and the number of topics covered in their courses?

—To what extent are teachers' opinions about how mathematics should be taught related to their reported practices?

—How closely do student reports match teacher reports of reform-oriented instructional practices?

—Is there a relationship between use of reform-oriented practices and student achievement in mathematics (overall and on each of the five content domains in TIMSS)?

We focus on the eighth grade for a few reasons. First, a richer and more consistent set of practice-related variables is available for eighth-grade teachers and students across the three waves of data collection. In addition, the 1999 TIMSS study did not include fourth-grade classrooms, making analysis of trends across time more difficult. Finally, preliminary analyses of the 2003 fourth-grade practice items yielded less consistent results than the eighth-grade data, both between teachers and students and among countries, complicating the construction of meaningful scales to measure practice. Nevertheless, many of our analyses could easily be adapted to the fourth-grade data.

This study is intended to shed light on instruction in the United States during a period of extensive mathematics reform. But our understanding of what is happening in the United States can be enhanced by an examination of instruction and achievement elsewhere, so we use the TIMSS data to explore relationships in three other countries—Japan, Singapore, and the Netherlands—that differ in important ways from the United States. They are not intended to be representative of any larger set of countries but are included only to provide a comparative context for understanding U.S. mathematics instruction and achievement. Japan is an especially high-achieving country, whereas achievement scores for students in the Netherlands are approximately midway between those of the United States and Japan. Both of these countries participated in the 1999 TIMSS video study, which provided rich information about classroom environments.[22] The video study suggests some potentially important differences among the three countries in how mathematics is taught. For example, classrooms in the Netherlands were characterized by a greater emphasis on real-life mathematics problems, more individual and small-group work, and greater use of calculators than in the United States. Japanese classrooms stood out from all of the other participating countries in several respects, including a larger amount of time devoted to the introduction of new content, a larger percentage of problems that were judged to be of high complexity, a larger percentage of problems that involved making connections, and a lower percentage of problems that were judged to be repetitions.[23] Singapore did not participate in the 1999 video study, but it is consistently one of the highest-scoring nations and its mathematics curriculum has been the subject of great attention among U.S. educators and policymakers.[24] Together, the four countries represent a wide variety of classroom contexts in which to examine the use of reform-oriented instructional practices.

Analytic Approach

As discussed earlier, the focus of this paper is on exploring the utility of the TIMSS survey data for understanding teachers' use of reform-oriented instructional practices and their relationships with achievement in mathematics. In this section we discuss the data, instruments, and statistical methods used to analyze the data.

Data

We rely on three primary sources of data: the eighth-grade mathematics teacher survey files from the 1995, 1999, and 2003 administrations of TIMSS in the United States, Japan, the Netherlands, and Singapore; the eighth-grade students' mathematics test-score files from the 2003 administration in each of these countries; and the 2003 student background survey data sets. We also refer briefly to other data sources, such as the video studies, where relevant. However, the focus of this paper is on the surveys rather than the videos because we are interested in examining the utility of survey-based measures for understanding instructional practice.

For the analyses of instructional practices over time, we use the teacher survey data sets for each of the four countries across the three waves of TIMSS. The surveys include questions about personal background, education, and experience, and also about teacher use of, or perceptions about, different kinds of instructional practices in the classroom. Our analyses of the relationship between instructional practices and student achievement refer to 2003 data only. For each country these analyses combine the student achievement data sets with those containing the information from the student and teacher background surveys.

TIMSS data have several advantages over other data sources, including the availability of information for a large number of countries, representative sampling of classrooms, and high-quality measures of student mathematics achievement. At the same time, these data have a number of limitations. Perhaps most significantly, the TIMSS tests were designed to ensure the validity of cross-country comparisons, not for sensitivity to instructional practice.

Coupled with differences in curriculum across countries, this could mean that for some of these countries the tests are less than optimally sensitive to instruction. Moreover, the TIMSS data sets do not allow researchers to follow individual students over time. It is widely acknowledged that the most valid approach for evaluating the effects of any educational intervention on student achievement (short of conducting a randomized experiment or quasi-experiment) is to measure each individual's achievement at several points in time to control for unmeasured student characteristics.[25] The absence of multiple measures of individual students' achievement over time severely limits the range of analyses that can be conducted and the strength of conclusions that can be drawn from these analyses. Additional limitations include a somewhat sparse set of survey items directly addressing reform-oriented instructional practices, changes in the wording of items from one wave to the next, and the strong possibility of incomparability of survey responses across countries. We address these and other limitations later in this paper.

Instructional Practices and Perceptions Items and Scales (2003 Student and Teacher Surveys)

Table 6A-1 in the appendix presents the complete list of items in the 2003 student and teacher surveys that address instructional practices in the classroom. To measure exposure to reform-oriented practices, we identified items on the teacher and student surveys that ask about practices consistent with a reform-oriented approach to mathematics instruction. We examined the item correlations and factorial structures separately for each country and estimated the internal consistency of the resulting scales.

Five items were included in the reform-oriented instruction teacher scale. They were listed in response to the question, In teaching mathematics to the students in the TIMSS class, how often do you usually ask them to do the following?:

—Work on problems with no immediately obvious method of solution
—Work together in small groups
—Relate what students are learning in mathematics to their daily lives
—Explain their answers
—Decide on their own procedures for solving complex problems

These items are similar to those used in other studies of reform-oriented practices, though there are fewer such items in the TIMSS surveys than in many of the other studies.[26] For each item teachers were asked to indicate whether they asked students to engage in each activity every or almost every lesson, about half the lessons, some lessons, or never. The internal consistency reliability of the composite created from these five items (using the average of the 1–4 response scale) is moderate (alpha = 0.64 for U.S. teachers and between 0.45 and 0.54 for the other three countries).[27]

Unlike some of the other studies examining reform-oriented instructional practices, we did not create a separate scale for *traditional* practices because the items in the TIMSS surveys do not lend themselves to such a scale. Instead, we created a content-focused composite to measure the frequency with which teachers asked students to engage in activities focused on the following activities:

—practice adding, subtracting, multiplying, and dividing without using a calculator

—work on fractions and decimals

—interpret data in tables, charts, or graphs

—write equations and functions to represent relationships

The first of these would generally be considered to be associated with a more traditional approach to instruction (though, as noted earlier, this type of skill building is not precluded by a reform-oriented approach), whereas the other three could be associated with either pedagogical style. All of these items were positively correlated; the composite (alpha = 0.54 in the United States; 0.42 to 0.62 in the other countries) is intended to control for exposure to mathematics content in our subsequent achievement models.

The student survey items addressing reform-oriented practices did not cluster as clearly or consistently as the teacher survey items. In particular, the items showed stronger relationships with one another in the United States than in the other three countries, and their correlations with other, non-reform-oriented items varied substantially across countries. These differences suggest that students in different countries might be interpreting the items differently. Therefore we did not create student-level composite scores using these items; instead, the models of student achievement presented later in this study include some of these individual items as predictors. However, in interpreting any relationships observed there, it should be remembered that the precise meaning of attending a classroom where students report more or less of a certain activity may not be the same across countries.

Finally, we examined a set of items addressing teachers' opinions and perceptions about reform principles in mathematics instruction (see table 6A-1). Although these items do not directly measure teachers' actions in the classroom, their content gets at the heart of many of the differences between the views of reform advocates and detractors. Table 6-1 shows the results of an exploratory factor analysis of these items (for teachers in the United States), separating a factor capturing *reform-oriented perceptions* (including, for example, items related to multiple representations or solutions) and a second factor capturing what we consider more *traditional* perceptions (such as memorization, algorithms). Table 6A-2 suggests there are some differences in the perceptions

Table 6-1. *Factor Analysis of Perceptions Items in the U.S. Teacher Survey*

Perception items	Factor loadings	
	1	2
More than one representation (picture, concrete material, symbols, and so on) should be used in teaching a mathematics topic	0.71	−0.07
Solving math problems often involves hypothesizing, estimating, testing, and modifying findings	0.76	−0.16
There are different ways to solve most mathematical problems	0.61	−0.43
Modeling real-world problems is essential to teaching mathematics	0.71	−0.16
Mathematics should be learned as sets of algorithms or rules that cover all possibilities	−0.06	0.78
Learning mathematics mainly involves memorizing	−0.18	0.69
Few new discoveries in mathematics are being made	−0.23	0.56

Source: TIMSS 2003 Teacher Questionnaire.

of teachers in other countries: thus for example Japanese teachers may perceive algorithms/rules (item b) as closer to *reform* principles, while Dutch teachers may perceive hypothesis testing (item c) as more *traditional.* As with instructional practices, however, there seem to be more similarities than differences in teachers' perceptions across countries. Consequently, we constructed *reform* and *traditional* composites from the original sets of perception items for teachers in all four countries.[28]

Instructional Practices and Topics Covered across Time (1995, 1999, and 2003 Teacher Surveys)

One of the goals of this study is to investigate whether use of certain instructional practices (specifically reform-oriented practices) saw any changes from 1995 to 2003 in the United States. As in 2003, in the two previous administrations of the TIMSS (1995, 1999) teachers were asked to report on their use of a number of instructional practices in the classroom. Unfortunately, however, the list of instructional practice items in the 2003 teacher survey differs considerably from that used in the 1995 and 1999 surveys. In all, only four instructional practice items remained consistent across all three waves of the TIMSS study. Three of these items are included in our reform-oriented instruction scale, and the fourth is one of the content-focused items discussed earlier. The precise wording of the items each year is presented in table 6-2.

For each year and country we estimate the frequency with which teachers reported using each of these four instructional practices in the classroom. For

Table 6-2. *Select Instructional Practice Items from the 1995, 1999, and 2003 Versions of the Eighth-Grade Mathematics Teacher Survey*

Survey year	Practice items
1995 and 1999	
Question	In your mathematics lessons, how often do you usually ask students to do the following? :
Survey items	
Computational skills	Practice computational skills
Problem solving	Work on problems for which there is no immediately obvious method of solution
Equations representation	Write equations to represent relationships
Explain reasoning	Explain the reasoning behind an idea
2003	
Question	In teaching mathematics to the students in the TIMSS class, how often do you usually ask them to do the following? :
Survey items	
Computational skills	Practice adding, subtracting, multiplying, and dividing without using a calculator
Problem solving	Work on problems for which there is no immediately obvious method of solution
Equations representation	Write equations and functions to represent relationships
Explain reasoning	Explain their answers

Source: TIMSS 1995, 1999, and 2003 Teacher Questionnaires.

any given country and year, the samples of teachers included in the TIMSS study are not representative of the population of teachers in that country. Instead, the teacher estimates are weighted proportionally to the students they teach; therefore the weighted estimates presented should be interpreted as referring to the reports of teachers of nationally representative samples of students.[29]

We finally analyze information included in the three TIMSS teacher surveys concerning the topics teachers covered during the school year. The goal of this analysis was to explore whether the number of topics covered in eighth-grade mathematics changed over the past decade. Comparisons across time are complicated here by differences in the number of mathematics topics listed in the survey in each study (thirty-seven in 1995, thirty-four in 1999, and forty-five in 2003) and also by unfortunate inconsistencies in the scales used to report coverage of topics (in 1999 and 2003 we cannot separate between topics not covered yet and not covered at all during the year, and in 2003 we cannot separate between *not taught* and *just introduced*). To facilitate relative comparisons across time, we obtain the average proportion of topics that teachers in a country reported covering at all each year. Furthermore, because we expect curricula to differ in algebra classes and general mathematics classes in the United States, we obtain separate estimates in 2003 for these two types of classes.[30]

Multilevel Modeling of Student Reports of Instructional Practice and Student Achievement

The TIMSS sample includes only two teachers per school in the United States, and only one teacher per school in the other three countries; this structure prevents estimation of three-level models where the student, teacher, and school levels can be modeled simultaneously. Given our focus on classroom practice, we thus employ a series of two-level hierarchical linear models (with students nested within teachers) to investigate the distribution of student reports of instructional practice across classrooms as well as the relationship between student achievement and use of reform instructional practices in the classroom.[31] Unlike a two-level structure with students nested within schools, this approach avoids aggregating information about practice available by teacher to the school level, and it also provides accurate estimates of variance at the student level.[32]

Estimation of all multilevel models is carried out using the HLM software package.[33] We first estimate unconditional multilevel models to partition the total variance in student reports of instructional practice into within- and between-classroom components (that is, to estimate the intraclass correlation, or ICC) separately for each of four countries and thirteen types of instructional practices (fifty-two models in total). Similarly, twenty-four unconditional models partition the variance in student mathematics achievement within and between classrooms, one for each of four countries, and six dependent variables (overall score, and scores on each of the five content strands in TIMSS). At the student level, the equation treats the achievement of student i in classroom j as a function of the classroom mean level of achievement (β_{0j}) and each student's deviation from that mean (ε_{ij}); the classroom mean in turn is modeled as a function of the grand mean score (γ_{00}) and a classroom-specific deviation (u_{0j}).

$$(6\text{-}1) \qquad\qquad MTOT_{ij} = \beta_{0j} + \varepsilon_{ij},$$

$$(6\text{-}2) \qquad\qquad \beta_{0j} = \gamma_{00} + u_{0j},$$

where ε, and u are assumed to be normally distributed, with mean zero and variances σ^2, τ_β respectively. Thus, τ_β and σ^2 reflect respectively the variance in student achievement within and between classrooms.

We then estimate *conditional* models to gauge the degree of relationship between instructional practices and student achievement, controlling for student and teacher background. At the student level, the covariates include gender, age, number of books and appliances at home (as proxies for socio-economic status), and home use of a language different from that of the test. In addition, three items are included in the student-level model to capture students' perceptions of the frequency of exposure to the following instructional practices in the classroom: lecture style presentations, working on problems independently, and use of calculators. At the teacher level, we include years of experience, highest level of education attained, and classroom size as controls. Finally, the teacher-level model includes the composites described previously to reflect teacher reform-oriented perceptions and use of reform-oriented instructional practices in the classroom.

The level 1 equation incorporates the student covariates $(X_{1\ldots p})$ as predictors of student achievement, so that β_{0j} is adjusted for student background (see equation 6-3). At the classroom level (level 2), these adjusted classroom means are modeled as a function of classroom and teacher predictors $(M_{1\ldots q})$, while u_{0j} represents residual classroom variance. Finally, the parameters representing the *effects* of student-level covariates $(\beta_{1j} - \beta_{pj})$ are held constant across schools (as depicted in equation 6-5).

(6-3) $$MTOT_{ij} = \beta_{0j} + \beta_{1j}X_{1ij} + \cdots + \beta_{pj}X_{pij} + \varepsilon_{ij},$$

(6-4) $$\beta_{0j} = \gamma_{00} + \gamma_{01}M_{1j} + \cdots + \gamma_{0q}M_{qj} + u_{0j},$$

(6-5) $$\beta_{1j} = \gamma_{10}.$$

Note that while these models control for a few important student and teacher covariates included in the background questionnaires, the nature of the data collected by TIMSS (that is, cross-sectional, nonexperimental design with no prior-year scores for students) does not support causal interpretations. For example, if a model shows a strong relationship between instruction and achievement, it is not possible to determine with certainty whether instruction affects student achievement, whether high-achieving students receive certain kinds of instruction, or whether both effects occur. Our analyses can thus only be interpreted as indicative of relationships between variables and perhaps point at interesting avenues for more rigorous causal investigations.

Considerations for Estimation with Weights and Plausible Values

In estimating models of student achievement in TIMSS, it is central to consider that each student is tested with a different set of items. While this approach provides more reliable and economic achievement aggregates by country, it also introduces a degree of uncertainty in estimating the achievement of individual students. To account for this uncertainty, TIMSS generates five scores for each student from a posterior distribution of *plausible scores,* as opposed to single maximum likelihood estimates.[34] The multilevel models of student achievement in this study take into account the variation introduced by the use of this multiple *plausible values* methodology: each of the models described above is estimated five times separately, one for each plausible score, and the five plausible parameters are then pooled together to estimate the final model parameters.[35] An adjusted standard error for each pooled parameter is estimated from the formula:

(6-6) $$SE = \sqrt{U^* + (1 + M^{-1})B_m},$$

where B_m is the variance of the five plausible parameters, U^* is the average estimated variance of these parameters, and $M = 5$ in this case. The resulting pooled parameter estimates and tests of significance take into account the measurement error introduced by the use of plausible values instead of single scores and are the equivalent of the jackknifed estimates used to generate descriptive statistics in TIMSS.[36]

Moreover, the multilevel model estimates presented in this study are weighted at the student level to take into account the unequal sampling probabilities introduced by the TIMSS sampling framework. As mentioned previously, TIMSS does not survey representative samples of teachers, and thus weighted model estimates involving teacher- or classroom-level effects are representative at the student level only (that is, they refer not to teachers directly, but to the teachers of nationally representative samples of students).[37]

Findings from Analyses of TIMSS Survey and Achievement Data

In this section we present the results of several sets of analyses of the survey and achievement data. We begin with descriptive information on changes in teachers' reported practices in the United States and elsewhere and then investigate how teacher reports of instructional practice relate to their perceptions about reform instruction and to student reports of practice in the classroom. Finally, we examine the extent to which these reported practices are associated with achievement.

How Has Teachers' Use of Reform Practices Changed over the Last Decade?

One way to understand the extent to which mathematics reforms took hold in U.S. classrooms during the 1990s is to examine changes in teachers' use of reform-oriented instructional practices over the course of the last decade. Of course, such changes cannot be assumed to represent direct effects of the reforms, but they can serve a useful illustrative purpose.

Figure 6-1 presents the average responses of eighth-grade teachers in each country to four items that were included in identical or very similar forms in all three waves. The first three (explain reasoning, problem solving, writing equations) correspond to a reform-oriented approach, whereas the last (emphasis on computational skills) is generally considered an indicator of a traditional approach to instruction (though it is important to note that it also indicates a particular content focus that might or might not be related to pedagogical style). Figure 6-1 shows sizable increases in the average frequency with which U.S. teachers reported using each kind of instructional practice between 1995 and 1999, whereas in most cases the use in other countries did not change substantially.[38]

Figure 6-1. *Frequency of Use of Various Instructional Practices, by Country and Year, 1995, 1999, 2003*[a]

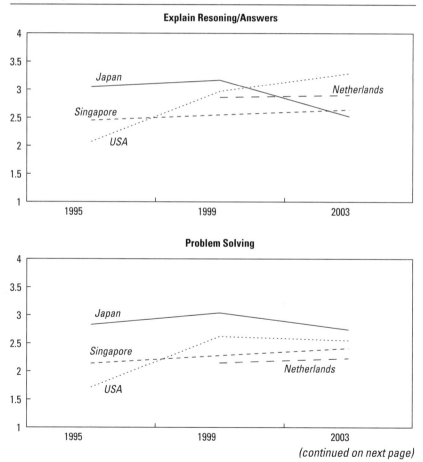

(continued on next page)

Figure 6-1. *Frequency of Use of Various Instructional Practices, by Country and Year, 1995, 1999, 2003*[a] *(Continued)*

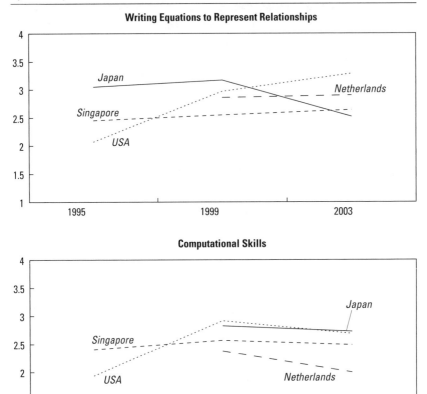

Source: TIMSS 1995, 1999, and 2003 Teacher Questionnaires.

a. Samples of teachers in each country are independent and different each year; some items in the teacher survey were not administered in the Netherlands and Japan in 1995; the scale ranges from 1 = never/almost never, to 4 = every lesson.

Between 1999 and 2003, reported frequency of explaining and problem solving continued to increase among U.S. teachers, while the reported emphasis on equations and computational skills remained relatively stable. As a result of the differences in trends among the four countries, the United States went from being the country in which the use of these practices was least frequent in 1995 to being among those where reported use was most frequent in 2003. While a causal interpretation of these summaries is clearly not possible, they suggest that something about the way mathematics was taught in the United States changed significantly during the 1990s. One possibility is that the changes reflect at least in part the influence of mathematics reforms under way in the country during the same period. At the same time, these trends suggest that increases in the reported use of instructional practices in the United States were not limited to reform-oriented items: as noted above, practice of computational skills also increased between 1995 and 1999 before leveling off.

Another way to examine changes in reform orientation is to consider the number of topics taught over the course of a year. Schmidt and colleagues have reported that mathematics curricula in the United States tend to include a relatively large number of topics, each of which is taught at a fairly shallow level.[39] Many reform-oriented mathematics curricula were designed to promote increased depth by focusing on fewer topics within each course than are typically covered in a more traditional curriculum, and by integrating key mathematical ideas into the curriculum throughout a student's mathematical education.

Here we present one analysis that is intended to shed some light on how breadth of coverage might differ across years and countries, but much more information would be needed to understand the nature of the mathematics content to which students are exposed. Moreover, although TIMSS gathered detailed information on topics taught each year, the inconsistencies among the surveys described previously potentially limit the extent to which it is possible to use the survey data to understand differences in depth and breadth across time, because trends in the data might reflect changes to the teacher survey in addition to (or instead of) real changes in breadth of coverage.

Table 6-3 presents the average proportion of topics (out of the total number listed in the survey) that teachers in each country reported spending any amount of time teaching during that year for each of the three TIMSS administrations. It points to a slight decrease in breadth for U.S. teachers from 1995 to 1999, a result that would be consistent with the changes in practice between 1995 and 1999 observed in figure 6-1 in that it suggests greater fidelity to reform principles. However, this trend is not as clearly reflected in the 2003 data.

Irrespective of trends over time, the overall results indicate that U.S. teachers reported teaching a higher average number of topics each year than did teachers in Japan, the Netherlands, and Singapore. At the same time, the table also indicates that this breadth of coverage does not apply to algebra classes, where U.S. teachers reported teaching about half as many topics as in U.S. general mathematics classrooms (a smaller proportion than in the Netherlands and Singapore and roughly the same as that of Japan). While it is difficult to draw strong conclusions from these results, they do suggest that the relatively high-achieving group of U.S. eighth graders who take algebra are exposed to instruction that is more closely aligned with reform principles in its focus on a smaller number of topics than is typical in U.S. mathematics classrooms.

Table 6-3. *Proportion of Topics in the Eighth-Grade Teacher Survey Actually Taught, by Year*

Country	Proportion of topics taught		
	1995	1999	2003
Japan	28.8	29.1	24.5
Netherlands	42.7	36.2	41.5
Singapore	38.9	24.4	39.2
United States	49.1	42.6	46.1 (overall)
			24.0 (algebra)

Source: TIMSS 1995, 1999, and 2003 Teacher Questionnaires.

Are Teachers' Perceptions Consistent with Their Practices?

Table 6-4 presents the correlations between composite indicators created after the factor analyses of teacher-reported practice and perception items described previously. Practice composites reflect emphasis on reform and content exposure, while perception composites reflect the extent to which teachers subscribe to traditional- and reform-oriented notions in mathematics instruction. In the United States traditional perceptions are negatively correlated to both reform perceptions (−0.23) and reform-oriented practice (−0.26), suggesting the two may indeed be perceived by teachers as being somewhat in opposition to one another. Teachers' use of reform-oriented practices is positively associated with their emphasis on mathematics content.

Interestingly, the positive correlation between reform practices and content is observed in the other three countries, but not the negative correlation between traditional and reform perceptions. While we have no information that would shed light on the source of these differences, it is possible that one result of the math wars in the United States in the 1990s was to make teachers perceive reform and traditional approaches as inconsistent with one another.

Do Student and Teacher Descriptions Match?

Because TIMSS 2003 gathered information about practices from both teachers and students, it is possible to compare the responses of these groups to evaluate the degree to which students' reports of instructional practices vary within and across classrooms and whether they match the reports provided by their teachers.

In comparing instructional practices across countries, it is informative to examine variability in student reports across classrooms (or similarly, the extent to which students in the same classroom provide consistent

Table 6-4. *Correlations among Teachers' Practices and Perceptions, by Country, 2003*

Country	Practice		Perceptions	
	Content	Reform	Traditional	Reform
Japan				
Content emphasis	1.00			
Reform practice	0.12	1.00		
Traditional perceptions	0.10	−0.01	1.00	
Reform perceptions	0.27*	0.20*	0.15	1.00
Netherlands				
Content emphasis	1.00			
Reform practice	0.32*	1.00		
Traditional perceptions	−0.16	−0.03	1.00	
Reform perceptions	0.18*	0.09	−0.08	1.00
Singapore				
Content emphasis	1.00			
Reform practice	0.41*	1.00		
Traditional perceptions	0.00	−0.07	1.00	
Reform perceptions	0.20*	0.27*	0.06	1.00
United States				
Content emphasis	1.00			
Reform practice	0.22*	1.00		
Traditional perceptions	0.08	−0.26*	1.00	
Reform perceptions	−0.01	0.33*	−0.23*	1.00

Source: TIMSS 2003 Teacher Questionnaire.
*$p < 0.05$.

information about instructional practices). Table 6-5 shows the interclass correlation in student reports of instructional practices for each of the four countries—that is, the proportion of variance between classrooms estimated from an unconditional multilevel model like that in equations 6-1 and 6-2. For all countries, between-classroom variation in the use of traditional or skill-focused practices is generally greater compared with practices we group under the reform-oriented label. In addition, the table points to a greater degree of between-classroom variability in student reports of instructional practices in the United States (and to a lesser extent in the Netherlands) compared with Japan and Singapore. While this measure is admittedly coarse, it suggests that the academic experiences of eighth-grade students may vary more considerably across mathematics classrooms in the United States, compared with students in other countries, who report more consistent instructional practices across classrooms. This result is consistent with the more heavily tracked nature of eighth-grade mathematics

Table 6-5. *Interclass Correlations for Student Reports of Instructional Practice, by Country, 2003*
Percentage of variance between classrooms

Instructional practice	Japan	Netherlands	Singapore	United States
Traditional				
a) We practice adding, subtracting, multiplying, and dividing without using a calculator	1.1	9.4	3.0	10.1
b) We work on fractions and decimals	2.5	3.6	4.7	10.3
i) We review our homework	19.9	38.4	8.2	34.4
j) We listen to the teacher give a lecture-style presentation	5.1	25.5	3.9	7.2
l) We begin our homework in class	9.9	12.7	12.5	38.1
m) We have a quiz or test	19.2	6.5	7.8	20.1
Average	9.7	16.1	6.7	20.1
Reform				
c) We interpret data in tables, charts, or graphs	1.4	6.4	3.0	14.8
d) We write equations and functions to represent relationships	3.8	9.8	9.5	14.5
e) We work together in small groups	22	38.2	20.9	39.5
f) We relate what we are learning in mathematics to our daily lives	3.7	6.7	5.1	11.2
g) We explain our answers	12.3	15.2	4.8	10.9
h) We decide our own procedures for solving complex problems	4.1	3.5	5.7	5.8
k) We work problems on our own	3.1	9.7	10.1	9.7
Average	7.2	12.8	8.5	15.2

Source: TIMSS 2003 Student Questionnaire.

education in the United States (as an example, the results in the previous section suggest that algebra classrooms are very different from regular mathematics classrooms).

Appendix table 6A-4 presents the average frequency of instructional practices reported by teachers and students. Compared with teachers, students across countries generally reported significantly more frequent use of traditional instructional practices or skills in the classroom (and of practices where they work without teacher supervision). By contrast, teachers tended to report more frequent use of reform-oriented instructional practices than their students. This could reflect teachers' better understanding of the nature and goals of different instructional practices, socially desirable answers, or a combination of both. Finally, table 6-6 presents the correlation between student reports of instructional practices and those of their teachers.[40] The results in the United States point to a small to moderate correlation between student and teacher reports of instructional practices, a weaker relationship than that reported in other studies.[41] In other countries the correlations are substantially lower, perhaps reflecting the limited variability of these classroom aggregates reported in table 6-6. In the absence of additional information with which to validate students' and teachers' reports, it is impossible to determine the source of these low correlations or to make conclusions about

Table 6-6. *Correlations between Student and Teacher Reports of Instructional Practices, by Country, 2003*

Instructional practice	Correlations			
	Japan	Netherlands	Singapore	United States
a) We practice adding, subtracting, multiplying, and dividing without using a calculator	0.00	−0.03	0.11	0.33
b) We work on fractions and decimals	−0.05	−0.06	−0.01	0.21
c) We interpret data in tables, charts, or graphs	−0.05	0.09	−0.04	0.22
d) We write equations and functions to represent relationships	0.07	0.18	0.15	0.23
e) We work together in small groups	0.24	0.25	0.14	0.55
f) We relate what we are learning in math to our daily lives	0.08	0.15	−0.01	0.26
g) We explain our answers	0.22	0.16	0.07	0.22
h) We decide our own procedures solving complex problems	0.04	0.12	0.15	0.19
k) We work problems on our own	−0.01	−0.01	0.06	0.02

Source: TIMSS 2003 Teacher and Student Questionnaires.

which source has greater validity. We include both in the multilevel models discussed in the next section.

Is Reform-Oriented Instruction Related to Student Achievement?

Table 6-7 presents the average mathematics scores of eighth-grade students in each country (total mathematics score and five subscores), along with the proportion of variance across classrooms estimated from the unconditional (or ANOVA) multilevel model of student achievement described in equations 6-1 and 6-2. In addition to the widely reported differences in the average achievement of students across countries, the table indicates that variation in mathematics achievement across classrooms is greatest in Singapore and the United States. Consistent with results reported in other studies, between half and three quarters of the total variance in mathematics scores in these countries is between classrooms.[42] By comparison, Japanese classrooms are much more homogeneous in terms of average mathematics achievement. Finally, 30–40 percent of variance in student scores lies between classrooms in the Netherlands.

The next modeling step is to investigate what characteristics of classrooms could help explain the greater variance in student mathematics scores in the United States and Singapore. These characteristics could include various kinds of resources, teacher training and background, and aggregate characteristics of the students, but they could also be related to variation in curricula and instructional practices. Tables 6-8 and 6-9 present the results of multilevel models of student achievement that investigated this question. As discussed before, the results cannot be interpreted in causal terms but merely as indicative of relationships that exist between student achievement and features of the classroom context in each country.[43]

Table 6-8 shows the results for the total mathematics score. Across countries the models consistently show a significant advantage for boys and for students of higher socioeconomic status (proxied through the number of books and appliances in the household) and significant disadvantages for nonnative speakers. There was also a negative relationship with student age (likely a proxy for grade retention) in the United States and the Netherlands, while the age coefficient was positive in Japan. In contrast, the relationship between student achievement and activities in the classroom (as reported by the students themselves) was inconsistent across countries. In the United States achievement was higher among students who reported working on problems on their own more often, and lower for students who reported listening to lecture-style presentations more often. The positive association with working on problems independently was also observed in other countries (and was strongest in Japan); however, lecture-style presentations were not significantly related to student achievement in other countries.

Table 6-7. *Average Student Mathematics Achievement and Interclass Correlations, by Country, 2003*

Category	Japan	Netherlands	Singapore	United States
Average achievement				
Total mathematics score	568.0	530.5	602.9	505.1
Data, analysis, probability	571.2	555.1	577.2	527.8
Fractions, number sense	554.5	533.1	615.1	508.4
Geometry	585.3	507.7	577.3	472.9
Algebra	566.1	508.5	587.2	510.6
Measurement	557.3	543.3	608.1	496.2
Proportion of variance between classrooms				
Total mathematics score	12.0	42.5	76.8	63.4
Data, analysis, probability	8.1	35.0	59.9	51.4
Fractions, number sense	11.7	41.1	73.6	62.1
Geometry	8.8	34.9	70.4	53.4
Algebra	10.1	39.5	65.9	59.3
Measurement	11.4	39.1	70.8	57.7
Sample sizes				
Number of students	4,147	2,397	5,214	6,841
Number of classrooms	125	103	283	346

Source: TIMSS 2003 Student Data files.

The effect sizes of these significant coefficients were often small, however.[44] For example, the model-based estimate of the standard deviation of total mathematics scores in the United States was 79.8; thus, a coefficient of 6.58 for *working on problems independently* represents only about 8 percent of a standard deviation. Even for students at the extremes of the four-point instructional practice scale, the effect size would be only about a third of a standard deviation.

Unlike the results at the student level, table 6-8 shows no significant association in the United States between student achievement and teacher background and experience, classroom size, and reform perceptions. For reform-oriented practice, the coefficient was positive but was not significant except in Singapore, where this indicator was positively and significantly related to achievement.

Table 6-9 presents the results of a similar model examining student achievement in algebra. Given the differentiation of curriculum in some U.S. classrooms,

Table 6-8. Multilevel Model of Student Achievement, Total Mathematics Score, Coefficient and Effect-Size Estimates, 2003[a]

	Japan		Netherlands		Singapore		United States	
Item	Coeff.	E.S.	Coeff.	E.S.	Coeff.	E.S.	Coeff.	E.S.
Classroom mean	567.74		533.00		602.61		504.52	
Teacher years of experience	−0.32	0.00	0.78	0.01	0.15	0.00	0.48	0.01
Teacher highest level of education.	15.26	0.20	14.03*	0.19	−0.31	0.00	0.26	0.00
Classroom size	0.86*	0.01	6.83*	0.09	0.99	0.01	−0.19	0.00
Teacher reform practice	7.49	0.10	−26.45	0.36	27.47*	0.14	12.31	0.16
Teacher reform perceptions	6.62	0.09	22.08	0.30	−11.33	0.34	−9.37	0.12
Boy	8.82*	0.11	13.08*	0.18	6.30*	0.08	7.48*	0.09
Different language at home	−5.35	0.07	−8.56*	0.12	2.55*	0.03	−5.31*	0.07
Books at home	9.35*	0.12	2.74*	0.04	0.31	0.00	7.26*	0.09
Lecture-style presentation	−0.23	0.00	1.03	0.01	0.26	0.00	−2.28*	0.03
Work on problems on own	31.96*	0.42	2.53	0.03	5.35*	0.07	6.58*	0.08
Use of calculators	−22.3*	0.29	0.19	0.00	−1.81	0.02	−1.77	0.02
Socioeconomic status (home appliances)	11.49*	0.15	1.52	0.02	1.91*	0.02	0.19	0.00
Student age	7.28*	0.09	−7.92*	0.11	0.60	0.01	−8.13*	0.10

Source: TIMSS 2003 Student and Teacher Data files.
*Statistically significant ($p < .05$).
a. The original practice scales were inverted in these models so that higher values represent more frequent use of instructional practice and the sign of the relationships is preserved.

Table 6-9. *Multilevel Model of Student Achievement, Algebra, Coefficient and Effect Size Estimates, 2003*[a]

Item	Japan		Netherlands		Singapore		United States	
	Coeff.	E.S.	Coeff.	E.S.	Coeff.	E.S.	Coeff.	E.S.
Classroom mean	565.76		511.24		586.94		509.75	
Teacher years of experience	−0.40	0.01	0.60	0.01	0.30	0.00	0.46	0.01
Teacher highest level of education	12.25	0.16	11.61*	0.15	1.61	0.02	−9.49	0.12
Classroom size	0.50	0.01	5.33*	0.07	0.59	0.01	−0.43	0.01
Time spent in algebra	−3.02*	0.04	4.28	0.05	−1.39	0.02	0.48	0.01
Teacher reform practice	47.63*	0.61	−25.17	0.32	74.45*	0.87	21.75*	0.29
Teacher reform perceptions	7.98	0.10	36.55	0.47	15.32	0.18	−10.83	0.14
Time in algebra by reform practice interaction	−1.20*	0.02	0.78	0.01	−1.30	0.02	−0.36	0.00
Boy	1.31	0.02	4.86	0.06	2.20	0.03	1.16	0.02
Different language at home	−6.11	0.08	−9.33	0.12	2.90*	0.03	−2.78*	0.04
Books at home	7.95*	0.10	3.14*	0.04	−0.03	0.00	6.70*	0.09
Lecture-style presentation	1.38	0.02	3.24	0.04	−0.78	0.01	−2.23*	0.03
Work on problems on own	27.77*	0.36	4.41	0.06	6.11*	0.07	5.64*	0.07
Use of calculators	−20.7*	0.27	−1.10	0.01	−2.38	0.03	1.14	0.01
Socioeconomic status (home appliances)	12.73*	0.16	7.90	0.10	2.18*	0.03	0.14	0.00
Student age	8.68*	0.11	−5.69*	0.07	4.62	0.05	−7.94*	0.10

Source: TIMSS 2003 Student and Teacher Data files.

*Statistically significant ($p < .05$).

a. The original practice scales were inverted in these models so that higher values represent more frequent use of instructional practice and the sign of the relationships is preserved.

this model included two additional predictor variables as controls: one was the proportion of time the teacher spent teaching algebra topics; and the second was the interaction between this indicator and the frequency with which the teacher reported using reform-oriented instructional practices in the classroom.

Except for the absence of a difference in the performance of boys and girls, the results at the student level shown in table 6-9 for algebra scores generally resemble those for the overall mathematics scores shown in table 6-9. In the United States, student achievement in algebra was positively related to working on problems independently (a relationship most strongly observed among Japanese students) and negatively related to lecture-style presentations (a relationship not observed in other countries). However, at least one potentially interesting difference was observed at the classroom level: unlike with total mathematics scores, teacher-reported frequency of reform-oriented practices was significantly and positively related to student achievement in algebra; moreover, in this case the associated effect size is more considerable at more than a quarter of a standard deviation. The same strong relationship between reform practices and student achievement was observed in Japan and Singapore. Finally, no significant relationships were observed between student achievement and time spent teaching algebra in the classroom or between the interaction of time in algebra and frequency of use of reform practices.

Appendix tables 6A-5 to 6A-8 present the results of multilevel models identical to that in table 6-9 for the relationships between student and classroom predictors and the other four mathematics subscores available in TIMSS (data, analysis, and probability; number sense; geography; and measurement). While the relevant coefficients were always positive, a constant in all these models was the absence of a significant relationship in the United States between student achievement and the frequency of so-called reform-oriented instructional practices reported by the teacher. In fact, the reform practice coefficient is only significant in the case of Singapore; in that country reform practice is significantly related to student achievement for all mathematics subscores.

Summary

Together, the analyses presented in this paper provide some evidence of changes in instructional practices that corresponded with the implementation of new mathematics curricula during the 1990s in the United States. Relevant findings include an increase in teachers' reports of the frequency with which they engaged in some instructional practices that are consistent with the mathematics reforms and a decrease in the number of mathematics topics presented over the course of a year. While the findings point to more stability in the practices reported by teachers in three other countries we investigated (Japan, the Netherlands, and Singapore), there is also the possibility that something about how teachers in the United States describe their practices changed over time while the practices themselves did not. The video studies conducted in 1995 and 1999 do not show an increase in reform-oriented instruction during that period, which suggests this last possibility should be given consideration.

Overall, the evidence of changes in U.S. teachers' use of reform practices is very limited and can only be interpreted as suggestive, among other reasons because of the limited number of practices compared. Even assuming the results reflect real changes in classroom practices, though, it is not clear what the implications of these changes might be for student achievement. The bulk of our cross-sectional analyses of 2003 student achievement data do not suggest a strong relationship in the United States between more frequent use of reform-oriented practices and higher student achievement.

Although it is hindered by a number of weaknesses in the design and analysis (including a design that does not support strong causal inference and a lack of consistency between student and teacher reports of practice), this study adds to a growing body of literature examining relationships between instruction and achievement. Understanding these relationships can help shape the mathematics education research agenda: A relationship that is observed in a variety of contexts would suggest the need for richer and better-designed research to understand the source of that relationship, whereas a consistent finding of no relationship might lead to the conclusion that research resources would be more wisely spent on other topics of research. Combined with the other studies reviewed earlier, the results presented here suggest that investing resources in promoting the kinds of instructional practices examined in this set of studies is unlikely to lead to detectable changes in student achievement. At the same time, the somewhat ambiguous results observed in this study (and others in the literature), combined with the limitations discussed in the next section, suggest that better methods for measuring instructional practice and different data collection designs are needed to understand more fully how the 1990s mathematics reforms influenced instruction and student achievement.

Limitations

The results of the analyses presented here must be interpreted very cautiously. The data and analytic approaches used have a number of limitations. Some of these are unique to TIMSS, but others characterize virtually all studies that examine instructional practices using survey data. We discuss four sets of limitations: validity problems with survey-based measures of instructional practice; the need to go beyond instructional practice to understand reform implementation; limitations inherent in the kinds of achievement measures used in large-scale projects like TIMSS; and limitations associated with the study design.

Survey-Based Measures May Not Accurately Capture Information about Practices

Perhaps the most significant problem in gauging the impact of reform-oriented instructional practice has to do with the specific aspects of reform that can be measured using instruments such as the TIMSS surveys. While the kinds of instructional activities examined in this paper are in many ways consistent with reform-oriented mathematics instruction, they by no means provide a complete picture, and in some instances might provide misleading information. In discussing the implementation of standards-based curriculum and instruction, Burrill acknowledges, "In some cases, there has been a focus on the 'trappings of the reform'—cooperative groups, manipulatives, hands-on activities—with little attention on mathematics as the focus of instruction."[45] Even when supplemented with information about curriculum or topics taught, most surveys fail to capture information about the extent to which teachers are implementing the core principles of reform, such as a focus on students' thinking. It is also difficult to create survey items that distinguish between different approaches to implementing a specific practice. The effectiveness of cooperative groups, for example, depends on how those groups are constituted and structured and on how the members interact.[46]

We also know from other research that teachers do not always interpret survey items in the ways the developers intend. Spillane and Zeuli used TIMSS survey items to identify high-reform teachers, but found that of the twenty-five teachers whose survey responses suggested high levels of reform practice, only four were judged by classroom observers to be teaching in a way consistent with reform ideals.[47] Hill's validation of instructional logs for elementary mathematics teachers showed that the accuracy of responses was affected by teachers' knowledge of terms and conventions, and interviews conducted by Le and others revealed that teachers sometimes mentally rephrased survey questions in ways that changed their meaning.[48] Low degrees of consistency between similar items on surveys and logs or between responses to the same survey administered at two time points also raise concerns about the validity of survey items.[49] However, Mayer found that consistency over time, as well as correlations between survey responses and observations, were relatively high when a composite of several items was used rather than a single item.[50]

The problem of differences in interpretation is likely to be especially acute for between-country comparisons. As shown earlier, factor analyses of student and teacher reports suggest that the instructional practice items do not always function consistently across countries, perhaps because of contextual differences involving language, curriculum, and other factors. Other research indicates that teachers in different countries often interpret survey questions differently, and differences have also been reported for student surveys.[51]

Another limitation of relying on teacher surveys is that they typically fail to capture information about differences in students' experiences within a classroom.[52] Teachers spend much of their time interacting with small groups or individuals, and even when engaged in whole-class activities might vary the instruction provided to different students through differences in the types of questions they ask or the feedback they provide. Individual students' experiences will also vary as a function of the characteristics and experiences they bring to the classroom, such as their level of engagement with the material.[53] It is impossible to fully understand cross-country differences in achievement without considering the many contextual and cultural factors that influence the academic experiences of students in classrooms.[54] Our analyses of student reports of instructional practice suggest that these experiences may vary considerably for students in different classrooms in the United States, while classrooms in Japan or Singapore tend to be more homogeneous. Also, exposure to content and instruction in earlier grades is likely to influence test scores and is an important aspect of opportunity to learn.[55] Thus a truly accurate measure of a student's exposure to instructional practices is very difficult to obtain with traditional survey methods and with the resource constraints typical of large-scale cross-sectional data collection efforts.

Finally, however closely we think teachers' or students' survey responses resemble their actual activities in the classroom, most surveys fail to provide any information on the *quality* of the practices in which teachers (and students) engage. There are a variety of ways to present complex problems or use cooperative groups, only some of which represent desirable instructional practice. This is an area where observational techniques such as those used in the TIMSS video studies can be especially valuable, but even these types of studies are likely to provide a less-than-complete picture of the quality of instruction to which students are exposed over the course of an academic year.[56] A range of promising new approaches for measuring instructional quality involving the use of different kinds of instructional artifacts are becoming available for researchers and are worth considering as ways to collect better information in large-scale studies.[57]

Data on Practices Provide Incomplete Information about Classroom Environments

Even if the validity of survey items were not a concern, it is clear that survey data about instructional practices provide only a limited understanding of what is actually happening in classrooms. Herman and Klein discuss three aspects of opportunity to learn, all of which can affect student performance: curriculum content, instructional strategies, and instructional resources.[58] Others argue that the cognitive demands of the curriculum are especially important and constitute a form of opportunity that should be examined.[59]

In particular, curriculum and instruction interact in complex ways to influence achievement, and an examination of one without any consideration for the other is likely to be incomplete. Evaluations of specific curricula often fail to take into account the ways in which reform-based curricula can become distorted in practice, so that the instruction students receive fails to match the goals of the curriculum developers. Similarly, evaluations of instructional practices should be informed by knowledge of what curriculum is in place. The relationship between practices and achievement has been shown in some cases to depend on the specific curriculum being implemented. McCaffrey and others found a relationship between reform practices and achievement in high school classrooms that were implementing reform-oriented mathematics curricula, but not in classrooms that were using more traditional textbooks.[60] Other studies have demonstrated the importance of considering pedagogy in the context of rigorous content when effects of instruction on achievement are examined.[61] To be of maximum utility, large-scale surveys should include detailed information both on practices and on curriculum, including the specific materials being used. Measuring the content of instruction is important for understanding how mathematics reforms have been implemented. Information on pedagogical strategies does not tell us, for example, whether teachers have integrated algebra into their instruction of early-grade students. But measuring content related to reform-oriented goals through surveys is not easy, and might be subject to even greater problems stemming from lack of shared understanding of terms than from measures of practice.[62] Although TIMSS questionnaires include some information about topics taught, they do not provide a sufficient level of detail to evaluate the implementation of core reform principles.

Some existing surveys provide more detail on instructional content. An example is the Surveys of Enacted Curriculum (SEC), which ask not only about content but also about the depth and rigor of that content.[63] Surveys like SEC are probably more likely to provide an accurate understanding of what is happening in the classroom but may pose a substantial response burden. Balancing the desire for accurate information about pedagogy, content, and depth with the need for efficient data collection is likely to remain one of the most significant challenges for studies like TIMSS.

Measures of Achievement Are Not Always Adequate for Supporting Intended Inferences

An additional concern stems from the need to consider the appropriateness of the achievement outcome for measuring instructional effects. Instructionally sensitive tests have the best chance of providing useful and valid information for educational improvement and evaluation.[64] The complex design and purposes of TIMSS create additional challenges. As Linn notes, "The more specific the purpose, the more homogeneous the population of students, and the narrower the domain of measurement . . . the easier is the task of developing measures that will yield results that support valid interpretations and uses."[65] The TIMSS assessments, by contrast, epitomize a situation characterized by multiple purposes, heterogeneity of students, and broad domains of measurement. These assessments were not designed (and were never validated) to detect effects of instructional practice, but to provide efficient and reliable country-level estimates of student attainment.

The validity of the TIMSS tests as measures of the effects of instructional approaches depends in part on the degree to which they are aligned with the curriculum to which students are exposed. The importance of this alignment is evident in the finding that country rankings can change substantially when only certain items are included in the achievement measure.[66] In addition, there is evidence of cross-country differences in the psychometric properties of the tests.[67]

At a more general level, different tests, test contents, or test formats may be more sensitive to instruction than others; and even when a test is said to be *sensitive to instruction,* the actual *degree of sensitivity* could in fact differ across subsets of items on the test.[68] Multidimensional scaling methodologies like the rule-space model allow analysts to further decompose student test performance by linking items to the different skills or cognitive processes they entail.[69] While potentially useful for exploratory or diagnostic purposes, however, such techniques pose substantial challenges for the modeling of instructional effects, as they entail large numbers of measures (that is, cognitive skills) that potentially are also differentially sensitive to instruction. Furthermore, scores on these measures may not be sufficiently reliable to detect relationships with instructional practice.

Finally, it is important to note that in addition to concerns about instructional sensitivity, the very nature of reform-oriented instruction can pose new and more fundamental challenges for measuring student achievement. As Webb has noted, for example, while group work can help promote collaboration and shared understanding in the classroom, it is crucial to determine the extent to which (or the conditions under which) it can also be expected to foster individual achievement.[70] These kinds of considerations have important implications for curricular design and for selection of the most appropriate approach to measuring student achievement (that is, individual- versus group-administered assessments).

From this it follows that modeling the relationship between student achievement and instructional practice may be especially difficult with TIMSS data;

not only are the available measures of instruction coarse (they were also designed with descriptive aggregates of teacher practice by country in mind), but it is reasonable to suspect that TIMSS achievement measures may be less than optimally sensitive to instruction.

Finally, exclusive reliance on multiple-choice items can produce an incomplete picture of student attainment in mathematics, with a relatively smaller weight assigned to dimensions of attainment more likely to be influenced through reform-instruction. Researchers have suggested that open-ended assessments or assessments that emphasize problem solving, understanding, and application may be more likely to demonstrate a positive relationship with reform-oriented instructional practices.[71] These results emphasize the importance of examining differences in relationships across outcome measures.

Large-Scale Survey Research Designs Limit Causal Conclusions

One of the primary limitations of most research conducted with large-scale databases is the inability to use research designs that support strong causal inference. In the absence of random assignment of teachers to the use of reform practices (or, perhaps more plausibly, to reform-oriented curricula), there is a high likelihood that teachers' use of those practices will be associated with other, unmeasured teacher characteristics and experiences. This confounding hinders our ability to make causal claims about the effects of reform practices on student achievement. This problem is especially severe in studies such as TIMSS that cannot follow individual students over time. Without the ability to examine previous achievement and previous exposure to reform practices, which could support a strong quasi-experimental design, it is impossible to conclude with any certainty that a correlation reflects a causal relationship.

What's Next?

After we reviewed existing literature and presented results of additional analyses, the answer to the question posed in the title of this paper—What can TIMSS tell us about mathematics reforms of the 1990s—could be summed up as "not very much." Although TIMSS and other international studies have provided valuable information on a range of topics, it is difficult to find much evidence in the data to support or refute the variety of claims that have been made about 1990s mathematics reform. Nonetheless, the ongoing TIMSS data collection provides an opportunity to enhance our understanding of the role that instructional practices play in influencing student achievement in mathematics, and given the tremendous resources currently devoted to international surveys, it is worth thinking about ways to make the most effective use of the data these surveys generate.

Although valuable information is obtained from the video studies, paper-and-pencil (or eventually on-line) surveys will continue to be relied upon for large-scale data gathering. The kinds of items included in the TIMSS surveys continue to dominate large-scale data collection efforts, but initiatives are under way to improve the quality of these measures and to develop measures of other important constructs such as teacher knowledge (the work of the Study of Instructional Improvement stands out).[72]

A few specific directions are especially important for researchers and developers of international education surveys to consider. The discussion in the previous section suggests that measures of general pedagogical strategies, such as whole-class or small-group instruction, should be supplemented with information about the content of the lessons taught, such as how algebraic concepts are integrated into instruction throughout the elementary grades. Valid measures of content probably need to include data collection activities other than surveys that ask teachers to report on the topics they taught. Classroom observations, examination of curriculum materials, and collection of artifacts such as lesson plans or classroom assessments are some examples of strategies that have been used to gather this information. These approaches vary in cost, feasibility, and fidelity to what is actually happening in classrooms.

Another important consideration for data collection involves whether to gather information about individual students' opportunities to learn. As discussed earlier, even within the same classroom, students' experiences may differ dramatically as a result of a number of factors: their own engagement and previous levels of knowledge and understanding, the attributes their peers bring to the class (which might be especially relevant when small-group instruction is used), and teachers' actions. This type of information could be collected from teachers (for example, by asking teachers to report on the instruction provided to an individual child rather than to the whole class) or students, and could also be gathered through observations and examination of student work.

A final issue, which has not yet been discussed in this paper, involves the role of external, contextual factors such as the existence of a national curriculum or high-stakes testing policies. TIMSS gathers extensive data on some of these contextual factors, and this information has contributed to our understanding of cross-nation differences in achievement. For studies focusing on mathematics education within the United States, contextual information collected at the state and district levels could enhance our understanding of classroom practices and student achievement. For example, states vary in the rigor and clarity of their content standards, in the difficulty and item formats of their state tests, and in the kinds of consequences they attach to test scores. Within each state, districts vary on such dimensions as the degree to which they mandate specific programs and the kinds of supports they provide for implementation of curriculum and standards. By gathering information on the various supports and sources of pressure affecting teachers' work, it might be possible to attain a more refined understanding of the interactions among curriculum, instruction, and achievement.

In the end, large-scale survey methods such as those used in TIMSS are inherently limited; they do not lend themselves to research designs that are best equipped to support causal inference, and they cannot possibly provide all of the contextual information needed to understand the mechanisms through which instructional practices influence achievement. But they provide an unparalleled opportunity to gather information across a wide range of districts, states, and countries, and therefore can play an important role in efforts to build the scientific research base in education. It is likely that such surveys will be with us for some time, and recent research provides a number of lessons that can be applied to survey development for future rounds of TIMSS and other large-scale studies.

Table 6A-1. *Instructional Practice and Perceptions Items in the 2003 TIMSS Student and Teacher Surveys*

Student Survey

Practices:
How often do you do these things in your mathematics lessons? (*Never* to *Every* or *almost every lesson*)
    a) We practice adding, subtracting, multiplying, and dividing without using a calculator
    b) We work on fractions and decimals
    c) We interpret data in tables, charts, or graphs
    d) We write equations and functions
    e) We work together in small groups
    f) We relate what we are learning in mathematics to our daily lives
    g) We explain our answers
    h) We decide on our own procedures for solving complex problems
    i) We review our homework
    j) We listen to the teacher give a lecture-style presentation
    k) We work problems on our own
    l) We begin our homework in class
    m) We have a quiz or test
    n) We use calculators

Teacher Survey

Perceptions:
To what extent do you agree or disagree with each of the following statements?
    (*Disagree a lot* to *Agree a lot*)
    a) More than one representation (picture, concrete material, symbols, etc.) should be used in teaching a mathematics topic

Table 6A-1. *Instructional Practice and Perceptions Items in the 2003 TIMSS Student and Teacher Surveys (Continued)*

b) Mathematics should be learned as sets of algorithms or rules that cover all possibilities
c) Solving mathematics problems often involves hypothesizing, estimating, testing, and modifying findings
d) Learning mathematics mainly involves memorizing
e) There are different ways to solve most mathematical problems
f) Few new discoveries in mathematics are being made
g) Modeling real-world problems is essential to teaching mathematics

Practices:

In teaching mathematics to the students in the TIMSS class, how often do you usually ask them to do the following? (*Never* to *Every* or *almost every lesson*)

a) Practice adding, subtracting multiplying, and dividing without using a calculator
b) Work on fractions and decimals
c) Work on problems for which there is no immediately obvious method of solution
d) Interpret data in tables, charts, or graphs
e) Write equations and functions to represent relationships
f) Work together in small groups
g) Relate what they are learning in mathematics to their daily lives
h) Explain their answers
i) Decide on their own procedures for solving complex problems

Table 6A-2. *Factor Analysis of Math Teachers' Perceptions about Reform Principles, by Country*

	Japan		Netherlands		Singapore		United States	
	Reform	Tradition	Reform	Tradition	Reform	Tradition	Reform	Tradition
a) More than one representation should be used in teaching a mathematics topic	0.72	-0.04	0.68	-0.03	0.68	0.07	0.71	-0.07
b) Mathematics should be learned as algorithms/rules that cover all possibilities	0.55	0.13	-0.12	0.75	0.40	0.55	-0.06	0.78
c) Solving math problems involves hypothesis, estimating, testing, modifying findings	0.67	0.11	0.13	0.66	0.67	0.21	0.76	-0.16
d) Learning mathematics mainly involves memorizing	0.24	0.78	-0.44	0.50	-0.16	0.67	-0.18	0.69
e) There are different ways to solve most mathematical problems	0.65	-0.10	0.51	-0.09	0.64	-0.24	0.61	-0.43
f) Few new discoveries in mathematics are being made	-0.31	0.64	-0.53	0.38	-0.01	0.67	-0.23	0.56
g) Modeling real-world problems is essential to teaching math	0.53	-0.22	0.63	0.31	0.65	-0.07	0.71	-0.16
Cronbach Alpha	0.59	0.11	0.32	0.45	0.60	0.31	0.65	0.45

Source: 2003 TIMSS Teacher Surveys.

Table 6A-3. *Average Use of Instructional Practices Reported by Teachers, by Country and Year*[a]

	Mean (standard error)		
Instructional practice	1995	1999	2003
Computational skills			
Japan	. . .	2.83	2.73
		(0.07)	(0.10)
Netherlands	. . .	2.38	2.00
		(0.08)	(0.07)
Singapore	2.41	2.57	2.49
	(0.09)	(0.08)	(0.04)
United States	1.94	2.92	2.69
	(0.09)	(0.07)	(0.05)
Problem solving			
Japan	2.18	2.48	2.49
	(0.05)	(0.07)	(0.05)
Netherlands	. . .	2.24	2.10
		(0.09)	(0.04)
Singapore	1.80	1.96	1.99
	(0.05)	(0.06)	(0.02)
United States	1.36	2.06	2.29
	(0.05)	(0.04)	(0.04)
Write equations to represent relationships			
Japan	2.83	3.04	2.74
	(0.06	(0.05)	(0.04)
Netherlands	. . .	2.15	2.23
		(0.06)	(0.05)
Singapore	2.14	2.28	2.41
	(0.05)	(0.05)	(0.03)
United States	1.71	2.62	2.55
	(0.08)	(0.04)	(0.04
Explain/reasoning answers			
Japan	3.05	3.17	2.52
	(0.06)	(0.06)	(0.05)
Netherlands	. . .	2.86	2.90
		(0.09)	(0.09)
Singapore	2.45	2.55	2.64
	(0.05)	(0.06)	(0.04)
United States	2.07	2.97	3.29
	(0.09)	(0.05)	(0.04)

Source: 1995, 1999, and 2003 TIMSS Teacher Surveys.

a. The statistical significance of the difference between two estimates across countries or years (Q1 − Q2) can be assessed through the formula $\left(Q_1 - Q_2 \right) \pm 1.96 \sqrt{ SE_1^2 + SE_2^2 }$. The difference is significant if the confidence interval around the difference does not contain zero.

Table 6A-4. *Mean Student and Teacher Reports of Instructional Practices, by Country*

Instructional practice	Japan		Netherlands		Singapore		United States	
	Student	Teachers	Student	Teachers	Student	Teachers	Student	Teachers
Practice add, subtract, multiply, divide without using a calculator	2.98	2.73*	2.72*	1.97	2.84*	2.51	2.92*	2.70
Work on fractions and decimals	2.47	2.03*	2.46*	1.99	2.75*	2.36	2.96*	2.57
Interpret data in tables, charts, graphs	2.57	2.42*	2.55*	2.35	2.37*	2.11	2.74*	2.26
Write equations and functions	2.67	2.73	2.61*	2.20	2.76*	2.39	3.08*	2.54
Work together in small groups	1.71	1.86*	1.75	2.23*	1.88	2.01*	2.32	2.60*
Relate learning in math to daily lives	1.97	2.04	2.00	2.34*	2.36	2.39	2.48	2.90*
Explain answers	2.42	2.52*	2.57	2.92*	2.82*	2.64	3.31	3.27
Decide own procedures for solving complex problems	2.32*	2.18	2.29*	2.06	2.55*	2.24	2.63	2.80*
Work problems on own	3.49*	2.50	3.52*	2.08	2.90*	2.00	3.38*	2.28

Source: 2003 TIMSS Teacher Surveys.

*Significant at $p < .05$. As before, the weighted student and teacher estimates were compared via a simple difference test with pooled standard error; see table 6A-3 for formula.

Table 6A-5. *Multilevel Model of Student Achievement: Data, Analysis, Probability, Coefficient, and Effect Size (E.S.) Estimates*

	Japan		Netherlands		Singapore		United States	
	Coeff.	E.S.	Coeff.	E.S.	Coeff.	E.S.	Coeff.	E.S.
Classroom Mean, G00	571.00	0.00	556.71		576.91		527.41	
Tchr. Yrs of Exp., G01	-0.26	0.00	0.61	0.01	0.15	0.00	0.35	0.00
Tchr. Highest Ed., G02	13.62	0.19	12.74*	0.18	-0.77	0.01	0.61	0.01
Classroom Size, G03	0.60	0.01	5.75*	0.08	1.14	0.01	0.03	0.00
Tchr. Reform Practice, G04	7.28	0.10	-19.98	0.28	21.63*	0.11	7.87	0.10
Reform Perceptions, G05	4.34	0.06	17.31	0.24	-8.51	0.27	-7.15	0.09
Boy, B2	11.38*	0.16	13.82*	0.19	11.60*	0.15	2.97	0.04
Diff. Language at Home, B3	-3.12	0.04	-8.40*	0.12	-1.78	0.02	-12.79*	0.17
Books at Home, B4	7.65*	0.11	4.58*	0.06	1.61	0.02	8.30*	0.11
Lecture-Style Present., B4	4.14	0.06	1.59	0.02	-0.71	0.01	-0.61	0.01
Work on Problems Own, B5	22.72*	0.32	1.72	0.02	2.99*	0.04	4.41*	0.06
Use of Calculators, B6	-22.1*	0.32	1.95	0.03	-1.25	0.02	-2.46	0.03
SES (Home appliances), B5	8.83*	0.13	6.33	0.09	2.40*	0.03	-0.09	0.00
Student Age, B1	5.98	0.09	-6.67*	0.09	-3.32	0.04	-7.84*	0.10

Source: 2003 TIMSS Teacher Surveys.
*Significant at $p < .05$.

Table 6A-6. *Multilevel Model of Student Achievement: Fractions, Number Sense Coefficient, and Effect Size (E.S.) Estimates*

	Japan		Netherlands		Singapore		United States	
	Coeff.	E.S.	Coeff.	E.S.	Coeff.	E.S.	Coeff.	E.S.
Classroom Mean, G00	554.18	0.01	535.66	0.01	614.80	0.00	507.89	0.01
Tchr. Yrs of Exp., G01	-0.44	0.20	0.73	0.19	0.16	0.00	0.48	0.01
Tchr. Highest Ed., G02	17.12	0.01	13.22*	0.09	0.06	0.00	0.60	0.01
Classroom Size, G03	1.04*	0.10	6.73*	0.33	1.04	0.01	-0.14	0.00
Tchr. *Reform* Practice, G04	9.13	0.08	-23.58	0.26	25.16*	0.13	11.25	0.14
Reform Perceptions, G05	7.38	0.08	18.80	0.22	-10.44	0.32	-9.92	0.12
Boy, B2	12.41*	0.14	15.85*	0.13	4.63*	0.06	10.07*	0.13
Diff. Language at Home, B3	-7.30	0.08	-9.49*	0.03	3.29*	0.04	-3.84	0.05
Books at Home, B4	10.79*	0.12	1.96	0.01	-0.42	0.01	7.59*	0.09
Lecture-Style Present., B4	-2.37	0.03	0.89	0.03	1.31	0.02	-2.14*	0.03
Work on Problems Own, B5	35.11*	0.40	2.28	0.01	5.89*	0.08	7.29*	0.09
Use of Calculators, B6	-22.4*	0.26	0.99	0.04	-1.96	0.03	-3.29*	0.04
SES (Home appliances), B5	12.34*	0.14	-2.84	0.04	2.13	0.03	1.47*	0.02
Student Age, B1	8.74*	0.10	-6.20*	0.09	-2.15	0.03	-8.46*	0.11

Source: 2003 TIMSS Teacher Surveys.
*Significant at $p < .05$.

Table 6A-7. Multilevel Model of Student Achievement (Geometry) Coefficient and Effect Size (E.S.) Estimates

	Japan		Netherlands		Singapore		United States	
	Coeff.	E.S.	Coeff.	E.S.	Coeff.	E.S.	Coeff.	E.S.
Classroom Mean, G00	584.74	0.00	508.34	0.01	577.07	0.00	472.59	0.00
Tchr. Yrs of Exp., G01	-0.19	0.00	0.78	0.17	0.23	0.00	0.32	0.00
Tchr. Highest Ed., G02	13.08	0.17	13.25*	0.17	0.59	0.01	-0.28	0.00
Classroom Size, G03	0.68	0.01	6.82*	0.09	0.73	0.01	-0.24	0.00
Tchr. Reform Practice, G04	4.61	0.06	-27.76	0.37	27.22*	0.11	10.42	0.14
Reform Perceptions, G05	5.53	0.07	19.41	0.26	-9.16	0.34	-8.49	0.12
Boy, B2	3.21	0.04	8.69*	0.11	6.51*	0.08	7.46*	0.10
Diff. Language at Home, B3	-3.90	0.05	-7.34*	0.10	2.55*	0.03	-4.61*	0.06
Books at Home, B4	8.48*	0.11	2.78*	0.04	1.33	0.02	5.46*	0.07
Lecture-Style Present., B4	0.93	0.01	1.54	0.02	0.11	0.00	-0.88	0.01
Work on Problems Own, B5	33.13*	0.42	-1.52	0.02	5.31*	0.07	5.35*	0.07
Use of Calculators, B6	-22.4*	0.29	-1.03	0.01	-1.84	0.02	-0.32	0.00
SES (Home appliances), B5	11.46*	0.15	15.88*	0.21	0.25	0.00	1.19	0.02
Student Age, B1	7.77*	0.10	-7.80*	0.10	4.11	0.05	-5.82*	0.08

Source: 2003 TIMSS Teacher Surveys.
*Significant at $p < .05$.

Table 6A-8. Multilevel Model of Student Achievement (Measurement) Coefficient and Effect Size (E.S.) Estimates

	Japan		Netherlands		Singapore		United States	
	Coeff.	E.S.	Coeff.	E.S.	Coeff.	E.S.	Coeff.	E.S.
Classroom Mean, G00	556.93		545.73		607.79		495.54	
Tchr. Yrs of Exp., G01	−0.32	0.00	0.73	0.01	0.18	0.00	0.46	0.01
Tchr. Highest Ed., G02	14.84	0.21	13.02*	0.18	−0.49	0.01	1.88	0.02
Classroom Size, G03	0.73	0.01	6.51*	0.09	1.09	0.01	−0.19	0.00
Tchr. *Reform* Practice, G04	7.56	0.11	−21.51	0.29	25.39*	0.11	11.81	0.15
Reform Perceptions, G05	7.50	0.11	18.40	0.25	−8.65	0.32	−9.63	0.13
Boy, B2	5.69	0.08	18.97*	0.26	10.55*	0.13	13.55*	0.18
Diff. Language at Home, B3	−5.85	0.08	−9.08*	0.12	1.19	0.01	−5.20*	0.07
Books at Home, B4	7.73*	0.11	1.77	0.02	0.40	0.00	6.59*	0.09
Lecture-Style Present., B4	−1.87	0.03	−1.10	0.02	−0.77	0.01	−1.03	0.01
Work on Problems Own, B5	29.49*	0.42	6.55*	0.09	6.14*	0.08	4.34*	0.06
Use of Calculators, B6	−18.0*	0.25	−0.79	0.01	−2.94*	0.04	−1.80	0.02
SES (Home appliances), B5	10.10*	0.14	−0.69	0.01	0.84	0.01	−0.08	0.02
Student Age, B1	4.85	0.07	−6.69*	0.09	−1.44	0.02	−6.22*	0.08

Source: 2003 TIMSS Teacher Surveys.
*Significant at $p < .05$.

Notes

1. National Council of Teachers of Mathematics (NCTM), *Curriculum and Evaluation Standards for School Mathematics* (Reston, Va.: 1989); NCTM, *Principles and Standards for School Mathematics* (Reston, Va.: 2000); A. C. Porter and others, *Reform of High School Mathematics and Science and Opportunity to Learn* (Rutgers University, Consortium for Policy Research in Education, 1994); M. C. Linn and others, *Teaching and Learning K-8 Mathematics and Science through Inquiry: Program Reviews and Recommendations,* unpublished report commissioned by the North Central Regional Educational Laboratory, Naperville, Ill., 2000 (www.ncrel.org/engauge/resource/techno/k8.htm).
2. For example, see NCTM, *Curriculum Focal Points for Prekindergarten through Grade 8 Mathematics: A Quest for Coherence* (Reston, Va.: 2006).
3. On testing computational skills, see, for example, T. Loveless, "Trends in Math: The Importance of Basic Skills," *Brookings Review* 21, no. 4 (2003): 41–43.
4. V. Le and others, *Improving Mathematics and Science Education: A Longitudinal Investigation of the Relationship between Reform-Oriented Instruction and Student Achievement* (Santa Monica, Calif.: RAND, 2006).
5. J. Hiebert, "Relationship between Research and the NCTM Standards," *Journal for Research in Mathematics Education* 30, no. 1 (1999): 3–19.
6. T. Lewin, "Report Urges Changes in the Teaching of Math in U.S. Schools," *New York Times,* September 13, 2006 (www.nytimes.com/2006/09/13/education/13math.html?ref=education).
7. M. T. Battista, "Research and Reform in Mathematics Education," in *The Great Curriculum Debate: How Should We Teach Reading and Math,* edited by T. Loveless, pp. 42–84 (Brookings, 2001). For additional discussion and examples of how each side is often depicted, see Loveless, *The Great Curriculum Debate.*
8. NCTM, *Curriculum Focal Points for Prekindergarten through Grade 8 Mathematics.*
9. S. Cavanaugh, "NCTM Issues New Guidelines to Help Schools Home in on the Essentials of Math," *Education Week,* September 12, 2006 (www.edweek.org/ew/articles/2006/09/12/03 nctm_web.h26.html); Lewin, "Report Urges Changes in Teaching of Math in U.S. Schools."
10. J. Bransford, A. Brown, and R. Cocking, eds., *How People Learn: Brain, Mind, Experience, and School* (Washington: National Academy Press, 1999).
11. Battista, "Research and Reform in Mathematics Education," p. 46.
12. L. M. Desimone and others, *The Distribution of Teaching Quality in Mathematics: Assessing Barriers to the Reform of United States Mathematics Instruction from an International Perspective* (Washington: National Research Council, Committee on Techniques for the Enhancement of Human Performance, 2005); S. P. Klein and others, *Teaching Practices and Student Achievement: Report of First-Year Results from the Mosaic Study of Systemic Initiatives in Mathematics and Science,* MR-1233-EDU (Santa Monica, Calif.: RAND, 2000); Laura S. Hamilton and others, "Studying Large-Scale Reforms of Instructional Practice: An Example from Mathematics and Science," *Educational Evaluation and Policy Analysis* 25: 1–29.
13. J. K. Jacobs and others, "Does Eighth-Grade Mathematics Teaching in the United States Align with the NCTM Standards? Results from the TIMSS 1995 and 1999 Video Studies," *Journal for Research in Mathematics Education* 37 (2006): 5–32.
14. On frequency of reform-oriented instruction, see Hiebert, "Relationship between Research and the NCTM Standards"; and J. L. Ravitz, H. J. Becker, and Y. T. Wong, *Constructivist-Compatible Beliefs and Practices among U.S. Teachers* (Irvine, Calif.: Center for Research on Information Technology and Organizations, 2000). On implementation, see M. T. Battista, "Teacher Beliefs and the Reform Movement in Mathematics Education," *Phi Delta Kappan* 75, no. 6 (1994): 462–70; Battista, "Research and Reform in Mathematics Education"; and J. P. Spillane and J. S. Zeuli, "Reform and Teaching: Exploring Patterns of Practice in the Context of National and State Mathematics Reforms," *Educational Evaluation and Policy Analysis* 21 (1999): 1–27.

15. Desimone and others, "The Distribution of Teaching Quality in Mathematics."
16. Bransford, Brown, and Cocking, *How People Learn: Brain, Mind, Experience, and School.*; Cobb and others, "Assessment of a Problem-Centered Second-Grade Mathematics Project," *Journal for Research in Mathematics Education* 22, no. 1 (1991): 3–29; J. G. Greeno, A. M. Collins, and L. Resnick, "Cognition and Learning," in *Handbook of Educational Psychology,* edited by D. C. Berliner and R. C. Calfee, pp. 15–46 (New York: Macmillan, 1996); J. Hiebert and T. P. Carpenter, "Learning and Teaching with Understanding," in *Handbook of Research on Mathematics Teaching,* edited by D. A. Grouws, pp. 65–97 (Reston, Va.: NCTM/Macmillan, 1992); and T. Wood and P. Sellers, "Deepening the Analysis: Longitudinal Assessment of a Problem-Centered Mathematics Program," *Journal for Research in Mathematics Education* 28, no. 2 (1997): 163–86.
17. Hamilton and others, "Studying Large-Scale Reforms of Instructional Practice"; D. P. Mayer, "Do New Teaching Standards Undermine Performance on Old Tests?" *Educational Evaluation and Policy Analysis* 20 (1998): 53–73; R. Shouse, "The Impact of Traditional and Reform-Style Practices on Student Mathematics Achievement," in *The Great Curriculum Debate,* edited by Loveless, pp. 108–33; J. Smith, V. Lee, and F. Newmann, "Instruction and Achievement in Chicago Elementary Schools: Improving Chicago's Schools" (Chicago: Consortium on Chicago School Research, 2001); and H. Wenglinsky, "How Schools Matter: The Link between Teacher Classroom Practices and Student Academic Performance," *Educational Policy Analysis Archives* 10, no. 12 (2002) (epaa.asu.edu/epaa/v10n12/).
18. L. Burstein, Z. Chen, and K. S. Kim, *Analyses of Procedures for Assessing Content Coverage and Its Effects on Instruction,* CSE Technical Report 305 (Los Angeles: National Center for Research on Evaluation, Standards, and Student Testing, 1989); L. Burstein and others, *Validating National Curriculum Indicators* (Santa Monica, Calif.: RAND 1995); and A. C. Porter, "The Uses and Misuses of Opportunity to Learn Standards," *Educational Researcher* 24, no. 4 (1995): 21–27.
19. H. Kupermintz and others, "Enhancing the Validity and Usefulness of Large-Scale Educational Assessments: I. NELS:88 Mathematics Achievement." *American Educational Research Journal* 32 (1995): 525–54; L. S. Hamilton, "Gender Differences on High School Achievement Tests: Do Format and Content Matter?" *Educational Evaluation and Policy Analysis* 20 (1998): 179–95; and J. R. Lockwood and others, "The Sensitivity of Value-Added Teacher Effect Estimates to Different Mathematics Achievement Measures," *Journal of Educational Measurement* 44, no. 1 (2007): 47–67.
20. Le and others, "Improving Mathematics and Science Education."
21. See, for example, Spillane and Zeuli, "Reform and Teaching."
22. Analyses of the 1999 video study relied on videos collected in 1995 for Japan, and in 1999 for the other countries; see J. Hiebert and others, *Teaching Mathematics in Seven Countries: Results from the TIMSS 1999 Video Study,* NCES 2003-013 (Washington: U.S. Department of Education, National Center for Education Statistics, 2003).
23. Hiebert and others, *Teaching Mathematics in Seven Countries;* J. W. Stigler and others, *The TIMSS Videotape Classroom Study: Methods and Findings from an Exploratory Research Project on Eighth-Grade Mathematics Instruction in Germany, Japan, and the United States,* NCES 1999-074 (U.S. Department of Education, National Center for Education Statistics, 1999).
24. As with many of the curricula used in the United States, the curricula in these countries have been subject to varying descriptions that reflect different opinions on their "reformedness." Singapore, for example, is often praised for its basic skills-oriented curriculum (see *New York Times,* "Teaching Math: Singapore Style," September 18, 2006), but the website for the U.S. editions of the Singapore math curriculum materials indicates that the curriculum "encourages active thinking process, communication of mathematical ideas, and problem solving" (www.singaporemath.com/Primary_Math_Textbook_1A_U_S_EDITION_p/pmust1a.htm).
25. Charter School Achievement Consensus Panel, "Key Issues in Studying Charter Schools and Achievement: A Review and Suggestions for National Guidelines," NCSRP White Paper Series 2 (University of Washington, 2006).

26. For other studies or reform-oriented practices, see, for example, D. K. Cohen and H. C. Hill, "Instructional Policy and Classroom Performance: The Mathematics Reform in California," *Teachers College Record* 102, no. 2 (2000); 294–343; Le and others, "Improving Mathematics and Science Education."

27. For each country we also conducted separate exploratory factor analyses on the full set of practice items for students and teachers. The results using the student surveys were inconsistent across countries. The teacher survey results were more consistent, showing the five reform-oriented items clustering together in each country.

28. It should be noted, however, that the alpha reliability estimates for traditional perceptions were generally low across countries, indicating that these composites may be less consistently measured than would be desirable, and potentially compromising the meaningfulness of the composite.

29. M. O. Martin, *TIMSS 2003 User Guide for the International Database* (Boston College, 2005).

30. Algebra classes were defined as those for which teachers reported spending 75 percent or more of their instructional time on algebra topics. The survey item that enabled this distinction between algebra and general math classes exists only in the 2003 survey. That year the United States was the only country (of the four included in this study) where any eighth-grade classes met this criterion.

31. S. W. Raudenbush and A. S. Bryk, *Hierarchical Linear Models: Applications and Data Analysis Methods,* 2nd ed. (Thousand Oaks, Calif.: Sage, 2002).

32. M. C. Opdenakker and J. Van Damme, "The Importance of Identifying Levels in Multilevel Analysis: An Illustration of the Effects of Ignoring the Top or Intermediate Levels in School Effectiveness Research," *School Effectiveness and School Improvement* 11, no. 1 (2000): 103–130; J. F. Martinez-Fernandez, "A Multilevel Study of the Effects of Opportuity to Learn (OTL) on Student Reading Achievement: Issues of Measurement, Equity, and Validity," Ph.D. dissertation, University of California, Los Angeles, *Dissertation Abstracts International,* DAI-A 65/11. 3155006 (Ann Arbor, Mich.: University Microfilms International, 2005).

33. S. W. Raudenbush and others, *HLM6: Hierarchical Linear and Nonlinear Modeling* (Lincolnwood, Ill.: Scientific Software International, 2004).

34. See Martin, *TIMSS 2003 User Guide for the International Database.*

35. For a description of the procedures and formulas used to obtain these pooled estimates, see Raudenbush and others, *HLM 6: Hierarchical Linear and Nonlinear Modeling.*

36. J. D. Willms and T. Smith, *A Manual for Conducting Analyses with Data from TIMSS and PISA* (Montreal: United Nations Educational, Scientific, and Cultural Organization, Institute for Statistics, 2005) (www.unb.ca/crisp/pdf/Manual_TIMSS_PISA2005_0523.pdf).

37. See Martin, *TIMSS 2003 User Guide for the International Database.*

38. Standard errors for the yearly averages for each country, presented in appendix table 6A-3, range from 0.02 to 0.09. While providing a sense of the precision of the estimates, confidence intervals can produce overly conservative tests of significance; see N. Schenker and J. F. Gentleman, "On Judging the Significance of Differences by Examing the Overlap between Confidence Intervals," *American Statistician* 55 (2001): 182–86. Thus, the statistical significance of differences across countries and years can be assessed through a simple pooled standard error of differences test. While the specific critical value differs for each pair of estimates, differences in the 0.15 to 0.20 range and above are generally significant.

39. For example, see W. E. Schmidt, C. C. McKnight, and S. A. Raizen, *A Splintered Vision: An Investigation of U.S. Science and Mathematics Education* (Dordrecht, Netherlands: Kluwer, 1997).

40. For each instructional practice item, student reports of frequency were aggregated by classroom, and these classroom averages were correlated to those of the teacher.

41. See, for example, J. L. Herman and J. Abedi, *Issues in Assessing English Language Learners' Opportunity to Learn Mathematics,* CSE Technical Report 663 (Los Angeles: National Center for Research on Evaluation, Standards, and Student Testing, 2004); and J. L. Herman,

D. C. D. Klein, and J. Abedi, "Assessing Students' Opportunity to Learn: Teacher and Student Perspectives," *Educational Measurement: Issues and Practice* 19, no. 4 (2000): 16–24. Muthén and others found enough overlap between student and teacher reports to suggest that collecting information from both could be unnecessary; see B. Muthén and others, "Opportunity-to-Learn Effects on Achievement: Analytical Aspects," *Educational Evaluation and Policy Analysis* 17, no. 3 (1995): 371–403.

42. For other studies, see, for example, D. Koretz, D. McCaffrey, and T. Sullivan, "Predicting Variations in Mathematics Performance in Four Countries Using TIMSS," *Educational Policy Analysis Archives* 9, no. 34 (September 14, 2001) (epaa.asu.edu/epaa/v9n34/ [October 10, 2006]).

43. We also conducted several tests of the sensitivity of our inferences to model specification. In particular, we tested the effects of including the content emphasis items and found that these did not affect our conclusions about other predictors of student achievement; these predictors were thus dropped from the final models. We also examined the effects of using individual teacher practice items, rather than the composite scales, for the United States and found a consistent lack of relationship with student achievement.

44. We estimated overall *d* effect sizes by standardizing each parameter with respect to the pooled within- and between-classroom variance estimated in each multilevel model; see D. Cohen, *Statistical Power Analysis for the Behavioral Sciences,* 2nd ed. (Hillsdale, N.J.: Lawrence Erlbaum Associates, 1988). Coefficient *d* gives a general sense of the importance of the effects with respect to the scale of the dependent variable. Alternatively level-specific effect sizes can be estimated through the *proportion of variance* explained by a variable at each level in the model.

45. G. Burrill, "Mathematics Education: The Future and the Past Create a Context for Today's Issues," in *The Great Curriculum Debate,* edited by Loveless, p. 37

46. J. R. Anderson, L. M. Reder, and H. A. Simon, "Applications and Misapplications of Cognitive Psychology to Mathematics Education," *Texas Educational Review* (Summer 2000) (act-r.psy.cmu.edu/papers/misaplied.html); and D. Druckman and R. A. Bjork, eds., *Learnng, Remembering, Believing: Enhanced Human Performance* (Washington: National Research Council, Committee on Techniques for the Enhancement of Human Performance, 1994).

47. Spillane and Zeuli, *Reform and Teaching.*

48. C. Hill, "Content across Communities: Validating Measures of Elementary Mathematics Instruction," *Educational Policy* 19, no. 3 (2005): 447–75; Le and others, "Improving Mathematics and Science Education."

49. D. P. Mayer, "Measuring Instructional Practice: Can Policymakers Trust Survey Data?" *Educational Evaluation and Policy Analysis* 21 (1999): 29–45; J. L. Smithson and A. C. Porter, *Measuring Classroom Practice: Lessons Learned from the Efforts to Describe the Enacted Curriculum—The Reform Up Close Study* (University of Wisonsin, Consortium for Policy Research in Education, 1994).

50. Mayer, "Measuring Instructional Practice."

51. On teacher surveys, see W. Schulz, "Testing Parameter Invariance for Questionnaire Indices Using Confirmatory Factor Analysis and Item Response Theory," paper presented at the annual meeting of the American Educational Association, San Francisco, 2006; on student surveys, see M. Walker, "The Choice of Likert or Dichotomous Items to Measure Attitudes across Culturally Distinct Countries in International Comparative Educational Research," paper presented at the annual meeting of the American Educational Association, San Francisco, 2006.

52. A. Gamoran, "Schooling and Achievement: Additive versus Interactive Models," in *Schools, Classrooms, and Pupils: International Studies of Schooling from a Multilevel Perspective,* edited by S. W. Raudenbush and J. D. Willms (San Diego: Academic Press, 1991).

53. R. E. Floden, "The Measurement of Opportunity to Learn," in *Methodological Advances in Cross-National Surveys of Educational Achievement,* edited by A. C. Porter and A. Gamoran, pp. 229–66 (Washington: National Academy Press, 2002).

54. H. W. Stevenson, S. Lee, and J. W. Stigler, "Mathematics Achievement of Chinese, Japanese, and American Children," *Science* 231, no. 4739 (1986): 693–99.

55. Floden, "The Measurement of Opportunity to Learn."

56. J. Hiebert and J. W. Stigler, "A Proposal for Improving Classroom Teaching: Lessons from the TIMSS Video Study," *Elementary School Journal* 101 (2000): 3–20.

57. For example, see L. C. Matsumura and others, "Measuring Instructional Quality in Accountability Systems: Classroom Assignments and Student Achievement," *Educational Assessment* 8, no. 3 (2002): 207–29; B. M. Stecher and others, "Using Structured Classroom Vignettes to Measure Instructional Practices in Mathematics," *Educational Evaluation and Policy Analysis* 28 (2004): 101–23; M. A. Ruiz-Primo and M. Li, "On the Use of Students' Science Notebooks as an Assessment Tool," *Studies in Educational Evaluation* 30 (2004): 61–85; and H. Borko and others, "Artifact Packages for Measuring Instructional Practice: A Pilot Study," *Educational Assessment* 10, no. 2 (2005): 73–104.

58. J. L. Herman and D. Klein, *Assessing Opportunity to Learn: A California Example,* CSE Technical Report 453 (Los Angeles: National Center for Research on Evaluation, Standards, and Student Testing, 1997); see also E. H. Haertel, "Differential Prediction and Opportunity to Learn," paper presented at the 2003 meeting of the American Educational Research Association, Chicago.

59. See, for example, L. A. Shepard, "The Role of Classroom Assessment in Teaching and Learning," in *Handbook of Research on Teaching,* edited by V. Richardson, 4th ed., pp. 1066–1101 (Washington: American Educational Research Association, 2001).

60. D. F. McCaffrey and others, "Interactions among Instructional Practices, Curriculum, and Student Achievement: The Case of Standards-Based High School Mathematics," *Journal for Research in Mathematics Education* 32, no. 5 (2001): 493–517.

61. A. Gamoran and others, "Upgrading High School Math Instruction: Improving Learning Opportunities for Low-Achieving, Low-Income Youth," *Educational Evaluation and Policy Analysis* 19 (1997): 325–38; A. C. Porter, "The Effects of Upgrading Policies on High School Mathematics and Science," in *Brookings Papers on Education Policy,* edited by D. Ravitch, pp. 123–64 (Brookings, 1998).

62. Burstein and others, *Validating National Curriculum Indicators;* Hill, "Content across Communities."

63. See (www.ccsso.org/projects/Surveys_of_Enacted_Curriculum/7804.cfm [September 12, 2006]).

64. E. L. Baker and R. L. Linn, "Validity Issues for Accountability Systems," in *Redesigning Accountability Systems for Education,* edited by S. Fuhrman and R. Elmore, pp. 47–72 (Teachers College Press, 2004).

65. R. L. Linn, "The Measurement of Student Achievement in International Studies," in *Methodolgical Advances in Cross-National Surveys of Educational Achievement,* edited by A. C. Porter and A. Gamoran, pp. 27–57 (Washington: National Academy Press, 2002), p. 27.

66. Schmidt, McKnight, and Raizen, *A Splintered Vision.*

67. K. Ercikan and K. Koh, "Examining the Construct Comparability of the English and French Version of TIMSS," *International Journal of Testing* 5, no. 1 (2005): 229–66; A. Grisay and others, "Translation Equivalence across PISA Countries," paper presented at the annual meeting of the American Educational Research Association, San Francisco, 2006.

68. R. Hanson, R. McMorris, and J. Bailey, "Differences in Instructional Sensitivity between Item Formats and between Achievement Test Items," *Journal of Educational Measurement* 23, no. 1 (1986): pp. 1–12; L. S. Hamilton, D. F. McCaffrey, and D. M. Koretz, "Validating Achievement Gains in Cohort-to-Cohort and Individual Growth-Based Modeling Contexts," in *Longitudinal and Value-Added Models of Student Performance,* edited by R. Lissitz, pp. 407–35 (Maple Grove, N.M.: JAM Press, 2006).

69. On the rule-space model, see K. K. Tatsuoka, J. E. Corter, and C. Tatsuoka, "Patterns of Diagnosed Mathematical Content and Process Skills in TIMSS-R across a Sample of 20 Countries," *American Educational Research Journal* 41 (2004): 901–26.

LAURA S. HAMILTON AND JOSÉ FELIPE MARTÍNEZ

70. N. M. Webb, "Group Collaboration in Assessment: Multiple Objectives, Processes, and Outcomes," *Educational Evaluation and Policy Analysis* 17 (1995): 239–61.
71. See, for example, G. B. Saxe, M. Gearhart, and M. Seltzer, "Relations between Classroom Practices and Student Learning in the Domain of Fractions," *Cognition and Instruction* 17, no. 1 (1999): 1–24; Cohen and Hill, "Instructional Policy and Classroom Performance"; D. Thompson and S. Senk, "The Effects of Curriculum on Achievement in Second Year Algebra: The Example of the University of Chicago Mathematics Project," *Journal for Research in Mathematics Education* 32, no. 1 (2001): 58–84; Hamilton and others, "Studying Large-Scale Reforms of Instructional Practice"; and Le and others, "Improving Mathematics and Science Education."
72. See, for example, H. C. Hill, B. Rowan, and D. L. Ball, "Effects of Teachers' Mathematical Knowledge for Teaching on Student Achievement," *American Educational Research Journal* 42, no. 2 (2005): 371–406.

7

School Size and Student Achievement in TIMSS 2003

GABRIELA SCHÜTZ

S chool size is an interesting aspect of educational organization and has been a major topic of discussion in the past decades, both in academia and in politics, for two reasons. First, school size could have an impact on operational costs. Increased size might reduce redundancy and allow more resources to be bundled together at the individual school. This, in turn, could lead to cost savings and lower per pupil spending in larger schools. Second, the effect of size on a school's organizational structures or on the interactions among school members could have an impact on student achievement.

Although costs and student achievement are considerations for schools worldwide, most evidence on the relationship between school size and per pupil spending or student achievement comes primarily from the United States. By attempting to identify the association between school size and student achievement within other countries, the analysis presented here tries to fill this gap. However, issues relating to public spending or per pupil costs will not be discussed in this context since that requires more detailed knowledge of the cost structure of each educational system and of each individual school.

The author would like to thank Jens Ludwig, Joachim Winter, Ludger Wößmann, and participants at the Second International Research Conference of the International Association for the Evaluation of Educational Achievement for helpful comments and discussions. Any errors are those of the author. All results and regressions can be obtained from the author at schuetz@ifo.de.

Instead, the analysis focuses on the association between school size and student achievement as measured by the test scores obtained in the Trends in International Mathematics and Science Study (TIMSS). Although other school and student outcomes such as self-esteem, soft skills, and problem-solving competencies are certainly important, they are not discussed here since the focus of most countries' educational systems is on teaching the core subjects rather than the aforementioned competencies. In addition, such outcomes are more difficult to measure, which would make international comparisons even more problematic.

School size can be hypothesized to act upon student performance through several channels.[1] One of the most important factors is the influence of size on the curriculum and courses offered. Larger schools with more students in each grade are in a better position to offer several different courses targeted at students' individual needs or preferences. This enables instruction to be organized around more homogeneous study groups. However, there is a drawback to tailoring courses and their contents too much to the students' capabilities or preferences, since only a few students might want to elect the more demanding academic courses and thus might not acquire the necessary skills for a successful professional career. In this respect, the need for smaller schools to concentrate on the core curriculum might be an advantage even if it rules out the option to group pupils according to their abilities.[2]

Another channel through which school size might influence student learning is via the social interactions of the school members. Clearly, small schools offer better conditions for intensive social interactions between teachers and students as well as within these respective groups, thereby promoting a positive learning environment. On the other hand, larger schools can probably offer a greater number and variety of extracurricular activities. The participation of students in these activities can be seen as an important aspect of school life and might, via an increased sense of belonging, also act to foster student learning.

The possible effects of these two channels suggest that potentially, in terms of student achievement, middle-sized schools would perform best. Larger schools might have the advantage of allowing the formation of more homogeneous study groups and offering courses that are better tailored to students' needs. They also are likely to offer more extracurricular activities for their students. However, smaller schools might be better focused on the core curriculum and provide the advantage of quality of social interactions among their members. Given the advantages and disadvantages of small and large schools, it is possible that middle-sized schools can better balance these factors and that students in middle-sized schools therefore perform best. This kind of reasoning is in line with the results reported by Lee and Smith, which indicated that high schools enrolling between 600 and 900 students are most effective in

Figure 7-1. *U-Shaped and Inversely U-Shaped Relationship between School Size and Student Performance*

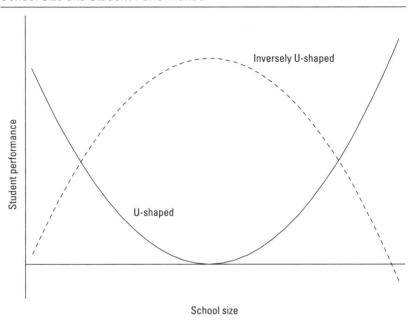

terms of student learning.[3] Based on their study, the relationship between school size and student performance can best be described as inversely U-shaped. This is illustrated in figure 7-1. With increasing enrollment levels, student performance at first increases until a certain threshold is reached, and thereafter the trend is reversed. Figure 7-1 also shows a U-shaped relationship, which implies that student performance at first declines with increasing enrollment levels and, after a certain turning point, improves with further increased enrollment.

However, a third factor might also influence the relationship between school size and student learning: the degree of autonomy of the individual schools. Within certain bounds, schools are free to organize their school life, courses, and teaching methods. This makes it possible for them to adapt their internal organization to their size. Moreover, in some cases and to some extent, schools also might be able to directly influence their size. Taking all these points into account, one might therefore also draw the conclusion that school size has no effect on student performance.[4]

The evidence on the effects of school size on student achievement is not conclusive.[5] However, some research suggests that middle-sized schools perform best and that student achievement is lower in very small and very large schools.[6]

This chapter uses TIMSS data to analyze the association between school size and student achievement for eighth graders within different countries. It also investigates whether there is such a thing as an optimal school size in an international context. Finally, it attempts to address the question of whether and how the relationship between school size and student achievement differs for different groups of students.

Data Set

This study uses student-level microdata from the 2003 Trends in International Mathematics and Science Study (TIMSS 2003) to analyze the association between school size and student achievement within different countries. TIMSS 2003 is a set of extensive international student achievement tests conducted by the International Association for the Evaluation of Educational Achievement, an independent cooperative of national research institutes and governmental research agencies. All participating countries received the same test items, so that the ensuing measures of the educational performance of students in math and science are directly comparable across countries.[7] Furthermore, test methodology provides for representative samples of students in each participating country.[8]

TIMSS 2003 target populations were students at the end of their fourth and eighth year of formal schooling in the participating countries. TIMSS 2003 was constructed as a curriculum-valid test, which ensures the validity of international comparisons of student achievement. To ensure comparability with previous TIMSS achievement tests, the formal definition for the grade eight student populations was the upper of the two adjacent grades with the largest share of thirteen-year-olds; this was usually the eighth grade in most countries.[9] At the eighth-grade level, data from forty-seven countries and four benchmarking participants are available for the analysis.[10]

TIMSS used item response theory scaling and calculated five plausible values for proficiency in mathematics and science for each participating student.[11] The scores were placed on the metric used to report results from previous assessments. The metric was established by setting the average of the mean scores of the countries that also had participated in TIMSS 1995 at the eighth-grade level to 500 and the standard deviation to 100.[12] The students' plausible values for mathematics are then used here as the measure of educational performance, while the science scores were only used to check the robustness of the results.

TIMSS provides not only achievement data for representative samples of students in the participating countries but also a rich array of background information on each student as well as his or her mathematics and science teachers and the school. In separate background questionnaires, students were asked to provide their gender, age, and information on various family characteristics, including the number of books in their home, their parents' education, and whether they themselves and their parents were born in the country where the test was taken. Background questionnaires were also answered by the students' teachers and the headmasters of the sampled schools. Thus additional information was gathered on the students' learning environment and the general background of their schools. Data from these questionnaires were combined using identifiers for the students, teachers, and schools. Since the analysis is based on the students' test performance in mathematics, only information on the mathematics teachers is included. In the cases where more than one teacher could be linked to a student, only information from the teacher who taught the most hours in the TIMSS class was included. If this did not uniquely identify a teacher (for example, due to missing information), the teacher who had the most teaching experience was chosen. This approach was adopted because if two teachers teach the same student, it can be hypothesized that the more experienced teacher is in a senior position to the less experienced one. In the cases where there was still more than one teacher to a student, one of the teachers was arbitrarily chosen to be included in the dataset.

The dataset contains data from fifty-one countries or regions. On average, the sample contains 141 schools in each country. Although schools were chosen at random, the design of the school sample in TIMSS was such that it provided an optimal sample of students and not of schools.[13] The sampling probability of a school was proportional to its size, which resulted in larger schools' being more frequently sampled than if a representative sample of schools had been drawn. Even though school size is analyzed only within the complete student sample and the sampling probability of each student is taken into account, one should bear in mind that the school sample is rather small and not representative for the population of schools. In some countries or regions, the number of schools sampled at the eighth-grade level was especially small. In Bahrain only sixty-seven schools were sampled; in Cyprus, fifty-eight; in England, fifty-seven; in Morocco and in Scotland, eighty-eight; and in the U.S. state of Indiana, only fifty.[14] Although results for these locations are presented, it should be borne in mind that inferences based on such small samples can be problematic.

With respect to the school size observed, in some of the participating countries, coding errors and outliers were identified and were accordingly omitted from the analysis.[15]

TIMSS used a two-stage stratified cluster design to sample students within each country. At the first stage, a sample of schools was drawn, and at the second stage, a sample of students was drawn out of this sample of schools.[16] Thus the primary sampling unit (PSU) in TIMSS was the school. The performance of students within the same school, however, may not be independent from one another.[17] This suggests that the independence assumption usually made with respect to individual observations in standard econometric methods should be relaxed in favor of the assumption that only the variation between schools (primary sampling units) provides independent variation. This is implemented by the clustering-robust linear regression method, which allows any given amount of correlation of the error terms within the same PSU and requires observations only to be independent across PSUs.[18] Since TIMSS used a stratified sampling design within each country, the sampling probabilities vary between students.[19] Nationally representative coefficient estimates are obtained by employing weighted least squares regression, using the sampling probabilities as weights. Weighted least squares estimation guarantees that the proportional contribution to the parameter estimates of each stratum in the sample is the same as if a complete census had been obtained.[20]

Framework of the Analysis

The framework of the present analysis is not only determined by the data and the method used but also by how school size is included in the education production function. Furthermore, as explained below, it is not possible to identify satisfactorily the causal effects of school size on student achievement with the data available. By including control variables, however, the relationship between size and test performance can be calculated net of other influencing factors.

Including School Size in Educational Production Functions

The easiest way to include school size in a standard educational production function is by treating it as a continuous variable and estimating a linear relationship. However, even at first sight, it is not clear why enrollment size should have a linear impact on student achievement. Increasing enrollment from 100 to 200 students probably has a different effect on social interactions and school organization than increasing enrollment from 1,000 to 1,100. Therefore, using nonlinear specifications with respect to school size seems more advisable. Moreover, the question of whether an optimal range of school size indeed exists cannot be addressed properly by including a linear term.

Two methods are capable of capturing the supposedly nonlinear and inversely U-shaped relationship between school size and student achievement. The first method entails constructing size categories and including dummies for these categories in the regressions. This method also has the advantage of imposing the least restrictions on the model. Relationships can be estimated much more flexibly by using size categories. With the TIMSS data, however, this approach was not possible. Due to the small sample of schools, the coefficients for the size category dummies could not be statistically significantly estimated in most cases.

The second possible approach, which is employed here, entails entering school size into the education production function together with its quadratic term. If an optimal school size range or an inversely U-shaped relationship indeed exists, this will be depicted by a positive coefficient on the linear term and a negative coefficient on the quadratic term.

Identifying Causal Effects and Unobserved Heterogeneity

It is not clear how the regression coefficients on size in an educational production function should reflect a true causal effect of size on student achievement. The problem encountered here, as in other similar analyses, is whether it can be reasonably assumed that children are randomly assigned to schools of different sizes.

This problem would be reduced considerably if one could argue that size is not a factor in the decisionmaking process when parents choose a school for their child. However, too little is known about the choice functions of parents and their children to support this view. If parents and students have the option to choose between different schools, they most certainly will do so. In most countries in the international dataset, parents have at least some freedom to choose the school for their child. The practice that residence determines which school the child will attend is not observed in most countries participating in TIMSS 2003. Furthermore, except in the United States and the United Kingdom, housing prices rarely reflect the quality of schooling. But even where choice is restricted, in most cases parents find ways to make the best possible choice for their child, especially when their socioeconomic status is higher.

Thus the problems arising from the selection of students into schools are apparent. If, for some reason, smaller schools are considered better in certain aspects of quality, all parents would try to send their children to the smaller (and supposedly better) schools. When choice is restricted or when capacity is a problem, better-off families are probably more successful in sending their children to the preferred schools. Since children from these families also perform better in student achievement tests, the results will be biased toward the finding that smaller schools are better.

One way to avoid this problem of nonrandom selection is to restrict the sample only to such schools and children that did not have the option to choose (either the school the students or the students the school). Unfortunately, this is not possible with the TIMSS 2003 data. Similarly, information identifying public and private schools is only available for a few countries.

The only strategy for dealing with nonrandom selection in the TIMSS data is to include as many control variables as possible in the regression equation to help reduce the selectivity bias. Including control variables will permit an estimate of the relationship between size and test performance net of other potentially influencing factors. These are factors that might affect the dependent variable (the test score) while being correlated with the explanatory variable of interest (school size). Not taking such factors into account will lead to biased estimates of the relationship between size and performance. If, for example, bigger schools reveal better student performance, it might be because of better-educated teachers or better resources in these schools. Also, schools of a certain size might attract a different student population, which is then reflected in differences in test scores. It is possible, for example, that large schools in cities attract more students with better-educated parents than do small schools in rural areas. This might result in a finding that larger schools are better whereas in fact they are only better because their students come from well-educated families. Also, if larger schools are better because they are better equipped, results will be biased in favor of finding that bigger is better if resources are not controlled for at the same time.

However, including control variables will only reduce the possible bias, not eliminate it. Therefore, unfortunately, regression coefficients cannot be interpreted as depicting causal effects. The use of the words "optimal" and "nonoptimal" in regard to school size and the associated highest or lowest predicted test scores should not be interpreted as a statement related to causality. Rather, these words are used to describe the kind of functional form found in the relationship between school size and test achievement.

Control Variables

Since TIMSS 2003 collected background information about the students, their families, and teachers and schools, it is possible to control for influencing factors at these levels. To control for student characteristics and family background effects, the regressions included student age; a dummy for student gender; three dummies for whether the father, the mother, and the student were born in the country where the test was taken; a dummy for whether the language of the test was sometimes or never spoken at home; the number of books at home in five categories; and a set of dummies for the highest level of education attained by the parents.[21] At the teacher level, the regression controlled for teacher experience and its square and for a set of dummies describing the teacher's level of education. Since the dependent variables are the students' plausible values in mathematics, the teacher variables referred to the students' teachers for this subject. Furthermore, at the school level, the regressions included the size of the TIMSS class and a dummy for whether more than 50 percent of the students enrolled at this specific school came from a disadvantaged background. It also included a set of dummies indicating whether the school principal reported that instruction in the school was affected by many or by some shortages in the following resources: instructional material, budget for supplies, school buildings and grounds, or teachers. Moreover, two dummies were included to capture the size of the community where the school was located. One describes whether more than 100,000 people live in the community and another whether fewer than 15,000 people live in the community. Since missing values had been imputed where possible, the regressions also included imputation dummies to control for imputation effects.[22]

Relationship between Size and Performance

The following analysis attempts to determine the kind of relationship that exists between school size and student performance in the countries participating in TIMSS 2003. Because of the suspected nonlinearity of the relationship, school size is included both as a linear and a quadratic term in the regressions. Control variables were added to capture the effects of school, teacher, and student background variables and to calculate the size effects net of these factors. Whether the relationship between school size and student performance differs for different groups of student populations is analyzed in a subsequent section. All regressions account for the stratified sampling design by employing student weights and clustering-robust linear regression.

Quadratic Relationship between School Size and Student Performance

Research on the effects of school size in the United States has sometimes found a kind of optimal size range, with small and large schools performing lower than middle-sized schools.[23] This sort of association would imply an inversely U-shaped relationship between school size and student performance. Therefore, the question of whether there exists an optimal school size was approached by including size as both a linear and a quadratic term in the regressions. If middle-sized schools reveal better student performance than larger or smaller schools, the regression coefficients on school size should be positive for the linear term and negative for the quadratic term.

Table 7-1 displays the results for the regressions that include school size and its square while at the same time controlling for family background and for teacher and school effects. The table contains the number of student observations available for the regression, the number of PSUs (that is, schools), and the estimated coefficients and standard errors for the linear and quadratic terms of school size. It also depicts the pattern of the relationship between test scores

Table 7-1. *Regression Describing Relationship between School Size and Student Performance*[a]
Units as indicated

Country or region	N	No. of schools (PSUs)	Coefficient on school size[b]	SE	Coefficient on school size squared[c]	SE	Pattern of relationship
Armenia	3,042	91	-1.202	4.007	0.078	0.234	…
Australia	3,288	143	2.895	3.444	-0.116	0.135	…
Bahrain	2,866	50	7.969	3.534**	-0.611	0.304*	Inversely U-shaped
Belgium (Flemish)[d, e]	4,788	139	2.990	3.668	-0.052	0.230	
Botswana[d]	4,662	132	7.625	5.037	-0.688	0.451	…
Bulgaria	3,124	132	-2.413	3.906	0.076	0.234	…
Chile[d]	6,264	192	3.324	1.603**	-0.072	0.055	…
Chinese Taipei	4,930	138	-0.421	1.157	0.006	0.020	…
Cyprus	3,291	50	10.203	7.926	-1.193	0.722	…
Egypt	5,085	156	0.391	1.831	-0.014	0.059	…
England	1,285	44	-6.319	13.548	-0.032	0.621	…
Estonia[d]	3,778	141	-0.296	2.478	0.113	0.176	…
Ghana[d]	2,808	113	9.705	5.353*	-0.452	0.609	…
Hong Kong, SAR	3,477	91	13.948	28.142	-0.351	1.288	…
Hungary	2,872	136	2.993	4.782	-0.295	0.414	…
Indonesia[d]	5,549	145	20.040	4.961***	-0.948	0.287***	Inversely U-shaped
Iran, Islamic Republic of[d]	4,863	178	4.042	4.826	-0.522	0.427	…
Israel	2,672	105	1.345	5.675	-0.040	0.331	…
Italy	3,690	151	-0.113	8.026	0.102	0.542	…
Japan[d]	4,835	145	-6.371	7.089	0.604	0.631	…
Jordan	3,962	125	5.434	3.576	-0.163	0.184	…
Korea, Rep. of	3,322	125	-0.336	2.634	0.097	0.109	…

Latvia	2,734	106	1.749	3.186	−0.078	0.209	⋯
Lebanon[d,f]	3,654	145	6.467	1.541***	−0.164	0.055***	Inversely U-shaped
Lithuania	3,620	114	2.705	2.567	−0.049	0.150	⋯
Macedonia, Rep. of	3,028	116	−0.750	3.893	−0.079	0.173	⋯
Malaysia	5,117	144	0.910	2.080	0.015	0.063	⋯
Moldova, Rep. of	2,075	80	−6.284	3.646*	0.285	0.208	⋯
Morocco[d]	1,986	88	1.373	1.414	−0.044	0.046	⋯
Netherlands	2,322	99	1.217	2.719	0.007	0.098	⋯
New Zealand	2,738	130	1.365	2.705	−0.059	0.115	⋯
Norway[d,g]	3,992	133	4.577	8.836	−0.386	1.241	
Palestinian National Authority[d]	3,757	118	−0.877	2.799	0.098	0.121	⋯
Philippines	5,891	116	0.690	0.694	−0.007	0.008	⋯
Romania	3,274	123	4.244	3.793	−0.143	0.174	⋯
Russian Federation[d]	4,571	210	−1.295	3.471	0.117	0.187	⋯
Saudi Arabia	2,925	118	2.826	4.899	−0.245	0.609	⋯
Scotland	1,629	65	−5.423	3.665	0.225	0.171	⋯
Serbia[d]	3,187	127	4.996	2.586*	−0.209	0.129	⋯
Singapore[d,h]	5,902	161	−22.645	4.880***	1.310	0.214***	U-shaped
Slovak Republic	3,558	152	−2.715	4.701	0.226	0.367	⋯
Slovenia[d]	3,178	154	−2.536	4.716	0.358	0.383	
South Africa[d]	7,366	211	7.563	2.471***	−0.303	0.158*	Inversely U-shaped
Sweden	3,172	123	3.410	5.212	−0.357	0.407	⋯
Syrian Arab Republic[d]	4,190	113	3.137	2.640	−0.185	0.109*	⋯
Tunisia[d]	4,594	140	−1.234	1.953	0.110	0.098	⋯
United States	6,117	173	−2.800	1.926	0.232	0.102**	⋯

(continued)

Table 7-1. *Regression Describing Relationship between School Size and Student Performance^a (Continued)*
Units as indicated

Country or region	N	No. of schools (PSUs)	Coefficient on school size^b	SE	Coefficient on school size squared^c	SE	Pattern of relationship
Benchmark regions							
Basque Country, Spain	1,902	87	3.677	1.676**	-0.103	0.060*	Inversely U-shaped
Indiana, United States^{d,i}	2,028	50	-3.715	5.547	0.034	0.269	. . .
Ontario, Canada	3,293	148	1.249	4.436	0.024	0.327	. . .
Quebec, Canada	3,173	127	-2.095	2.239	0.130	0.097	. . .

Source: Author's regressions based on TIMSS 2003 dataset. See Ina V. S. Mullis and others, *TIMSS 2003 International Mathematics Report: Findings from IEA's Trends in International Mathematics and Science Study at the Fourth and Eighth Grades* (Boston College, 2004).

*Significant at the 10 percent level; ** significant at the 5 percent level; *** significant at the 1 percent level.

a. Except where otherwise indicated, imputations at the school level were not possible and thus the regression includes imputed values at the student level only. Fewer than 100 schools were sampled in Bahrain, Cyprus, England, Morocco, Scotland, and in the U.S. state of Indiana. Inferences from such small school samples seem problematic. Moreover, the regression sample comprises fewer than 100 schools in Armenia, Hong Kong, Moldova, the Netherlands, and the Basque Country in Spain. The following countries and their states, provinces, or regions are members of the Organization for Economic Cooperation and Development: Australia, Belgium, Canada, England, Hungary, Italy, Japan, Republic of Korea, Netherlands, New Zealand, Norway, Scotland, Slovak Republic, Spain, Sweden, and the United States.

b. Coefficients multiplied by 100.

c. Coefficients multiplied by 10,000.

d. Regression includes imputed values both at the student and at the school level.

e. The regression in Belgium does not control for teacher education since there is no variation in this variable in Belgium.

f. In Lebanon, no information was provided on the education of the teachers. The regression therefore does not control for this variable.

g. In Norway, no information was provided on the percentage of students from disadvantaged family backgrounds. The regression therefore does not control for this variable.

h. The pattern of the coefficients in Singapore predicts not an optimal school size but rather a nonoptimal size where test scores are lowest.

i. In the U.S. state of Indiana, no information was provided on the experience of the teachers. The regression therefore does not control for this variable.

and enrollment as indicated by the estimated coefficients, that is, whether the relationship is U-shaped or inversely U-shaped.

Looking at the results of the regressions for each of the different countries contained in the dataset, an inversely U-shaped relationship between school size and student performance—and thus the existence of an optimal school size range—is not easily verified for most countries. As can be seen from the table, only six out of the fifty-one countries and regions demonstrate a significant quadratic relationship between school size and student test scores. An inversely U-shaped relationship between school size and student performance was found in Bahrain, Indonesia, Lebanon, South Africa, and in the Basque Country in Spain.[24] In contrast, Singapore showed a U-shaped relationship between school size and student performance. This suggests that initially test performance decreases as school size increases until a size threshold is crossed, at which point the trend reverses, becoming positive.

In the United States, the estimated coefficients on the quadratic and the linear term of school size suggest a positive relationship between school size and student performance over the entire school size range of between 38 and 2,489 students observed in the sample. As schools become larger, average test performance increases with increasing slope.

For the countries where the estimated coefficients on both the linear and the quadratic term of school size were statistically significant, the implied optimal or nonoptimal size can be calculated from the coefficients. This is done by applying simple algebra and calculating the level of enrollment for which the slope of the function between school size and student performance is zero. Depending on the pattern of the relationship, this enrollment level is either a maximum or a minimum. Table 7-2 displays this optimal or nonoptimal enrollment level together with the smallest and the largest school size in the regression sample for each country. Comparing the calculated optimums across the six countries in table 7-2 demonstrates that the optimal or nonoptimal size varies widely among the countries in the sample. The optimal size calculated for Bahrain is 652 students whereas for Lebanon an enrollment of 1,970 students seems to be optimal.

Comparing these optimums with the observed smallest and largest enrollment levels in each of the respective countries gives the impression that the calculated optimal or nonoptimal school sizes are located roughly around the middle of the school size distribution in the samples of students. This is important insofar as it implies that the results of the estimations are reasonably well nested within the data, providing confidence in the robustness of the relationship.

However, it is necessary to take a closer look at the distribution of school size within the student populations of the respective countries to be able to

Table 7-2. School Size Distributions in Countries with Significantly Estimated U-Shaped or Inversely U-Shaped Relationships
Units as indicated

Country or region	School size in regression sample (no. of students)		Optimal school size (no. of students)	Percentage of students at schools with			Implied change in test score points	
	Smallest	Largest		Enrollment higher than optimal size	Enrollment at 100 students > or < optimal size	Enrollment 100 students < optimal size	Enrollment increased from smallest to optimal	Enrollment reduced from optimal by 100 students
Bahrain	170	1,034	652	59.28	28.24		14.21	0.61
Indonesia	56	1,695	1,056	13.64	9.16		94.93	0.95
Lebanon	42	2,898	1,970	5.17	3.91		61.02	0.16
Singapore	197	1,852	864[a]	88.85	4.27		−58.33	−1.31
South Africa	68	1,647	1,247	12.96	8.54		42.15	0.30
Basque Country, Spain[b]	90	2,761	1,786	10.02	1.32		29.62	0.10

Source: Author's regressions and calculations based on TIMSS 2003 dataset. See also table 7-1.
a. For Singapore this value is a *nonoptimal* school size.
b. OECD member state.

judge whether the estimated relationship between size and performance is really plausible. A quadratic relationship might be fitted to the data while the result is in fact driven by just a few schools at either end of the size distribution. Therefore, the percentage of students within the respective countries that actually attended schools with enrollment levels higher than the suggested optimal or nonoptimal size is of interest. This figure suggests how much support for the quadratic specification is in the data.

In most of the countries where a quadratic relationship could be significantly estimated, the percentage of students attending schools larger than the estimated optimum is around 10 percent: 13.64 percent in Indonesia, 12.96 in South Africa, and 10.02 in the Basque Country in Spain. In Lebanon the percentage of students attending schools that are larger than the calculated optimum lies at 5.17 percent. Only in Bahrain (59.28 percent) and Singapore (88.85 percent) do more than 20 percent of the students in the sample attend schools with an enrollment level that ranges above the calculated optimal or nonoptimal size. Altogether, these figures support the view that the regression results are not driven by outliers.

Table 7-2 also shows the percentage of students attending schools that enroll between 100 students less and 100 students more than the suggested optimum. These numbers illustrate how dense the school size distribution is around the optimal or nonoptimal size. In all but one country, less than 10 percent of the student population attends schools with an enrollment that falls within this range. The exception is Bahrain, where 28.24 percent of the student population attends schools that enroll between 100 students less and 100 students more than the suggested optimum.

In Bahrain predicted test scores are highest at the calculated optimal school size of 652 students. This result is supported by the finding that 59 percent of the student population attends schools with enrollment exceeding the calculated optimum. This means that the calculated optimal school size lies well within the school size distribution observed in the student sample. The smallest school observed has 170 students while the largest has 1,034 students. An increase in size from the smallest enrollment (170 students) to the supposedly optimal level (652 students) increases test scores on average by 14 points. A reduction from the optimal level by 100 students leads on average to a loss in student performance of 0.61 test score points.

For Indonesia, the optimal school size is calculated at 1,056 students. Since school size in the regression sample varies between 56 and 1,695 students, the optimal school size seems to be located well within the sample. However, only 14 percent of the students attend schools that are larger than the optimum. The gain in test scores associated with increasing enrollment from the smallest school in the sample to one with an optimal enrollment level is 95 points.

If an optimally sized school experienced a reduction in size by 100 students, this would on average translate into a loss in student achievement of 0.95 test score points.

In Lebanon the estimated coefficients indicated an optimal school size of 1,970 students. The smallest school in the regression sample enrolls 42 students while the largest school enrolls 2,898 students. At the same time, the percentage of students that attend schools that enroll more students than the calculated optimum is relatively low at 5 percent. An increase in enrollment from the smallest to the optimal level is on average accompanied by a 61-point increase in test scores. Reducing an optimally sized school by 100 students would lead on average to a reduction of 0.16 test score points.

For Singapore a U-shaped relationship between school size and student performance was estimated. The results of the estimation imply that test scores are lowest in schools with an enrollment around 864 students. Students at both smaller and larger schools perform better. The smallest school in Singapore enrolls 197 students while the largest school enrolls 1,852 students. Approximately 89 percent of students in the sample attend schools that are larger than the estimated nonoptimal enrollment level. Thus, the nonoptimal enrollment level in this case is not very close to the middle of the students' distribution across differently sized schools. Increasing school size from the smallest school observed to the nonoptimal enrollment level on average leads to a reduction in achievement of 58 test score points. In comparison to the other five countries, the relationship between school size and student achievement is stronger. A reduction of the nonoptimal enrollment level by 100 would result in an increase in test scores by 1.31 points.

In South Africa, the smallest school size observed in the regression sample is 68 students while the largest school consists of 1,647 students. The optimal school size is estimated at 1,247 students, and 13 percent of the students in the regression sample attend schools that are larger. Increasing enrollment from 68 to the optimal level of 1,247 results on average in an increase in student performance by 42 test score points. At the same time, student achievement is reduced by only 0.30 test score points if the enrollment level at an optimally sized school is reduced by 100 students.

In the Basque Country in Spain, an enrollment level of 1,786 students seems to be optimal for student achievement. The smallest schools in the sample enroll only 90 students while the largest schools enroll as many as 2,761 students. However, in this case, the percentage of students attending schools that are larger than the supposed optimum is rather low at 10 percent. Moreover, the relationship between school size and student achievement is not very strong. Although test scores rise by 30 points on average when students from the smallest schools are compared to those from schools at the optimal level, the difference in student achievement between an optimally sized school and one with 100 students less is very low at 0.10 test score points.

Linear Relationship between School Size and Student Performance

For some countries the relationship between school size and student performance was better described as constant returns to size. School size was then included only as a linear term in the regression equation. Results for these regressions are depicted in table 7-3. The results show significant linear relationships between size and performance in fourteen of the fifty-one countries. The relationship is positive in eleven of these fourteen countries and negative in three.

A positive relationship between school size and student achievement is found in Belgium, Chile, Estonia, Ghana, Hong Kong, Jordan, Republic of Korea, Lithuania, Malaysia, the Netherlands, and Tunisia. The strength of the relationship varies considerably among those countries. An increase in school size by 100 students translates in most of these countries into an increase in test scores between 1.1 and 2.6 points. In contrast, the same increase in school size is associated with an increase in student achievement of 5.9 test score points in Ghana and 6.4 test score points in Hong Kong.

For England, Macedonia, and the U.S. state of Indiana, a negative relationship between school size and student achievement is found. In England an increase in size by 100 is on average associated with a 7-point decrease in scores. In Macedonia and Indiana, the same increase in size translates on average into a decrease of 2.5 and 3.1 test score points, respectively.

Thus the conclusion from the analysis is that the shape of the relationship between student performance and school size differs widely across the countries participating in TIMSS 2003. In some countries, the relationship seems to be positive over the whole range of school size observed, while in other countries it seems to be rather negative. In other countries, a quadratic relationship—either inversely U-shaped or U-shaped—seems to apply. And even if only those countries with an inversely U-shaped relationship are compared, the predicted optimal enrollment level still varies considerably. However, this is not really surprising since the school size range observed in the data also varies widely between the countries. The difference in enrollment between the largest and the smallest school in the sample is comparably low in Norway and Cyprus, where these differences amount to only 541 and 619 students. On the other hand, in countries like the Philippines and Chinese Taipei, the difference in enrollment between the largest and the smallest school is rather high, with differences of 9,245 and 4,783 students, respectively.

Table 7-3. *School Size Distributions in Countries with Significantly Estimated Linear Relationships*[a]
Units as indicated

Country or region	N	No. of PSUs	Coefficient on school size[b]	SE	School size (no. of students)		Implied change in test score points with enrollment increased from smallest to highest level observed
					Smallest in regression sample	Largest in regression sample	
England	1,285	44	−7.022	2.503***	416	1,807	−97.679
Indiana, U.S.[c,d]	2,028	50	−3.119	1.320**	49	1,800	−54.609
Macedonia, Rep. of	3,028	116	−2.452	1.034**	60	2,300	−54.930
Tunisia[c]	4,594	140	1.054	0.483**	33	1,991	20.634
Estonia[c]	3,778	141	1.326	0.763*	82	1,359	16.939
Malaysia	5,117	144	1.369	0.623**	220	3,296	42.102
Netherlands	2,322	99	1.412	0.834*	175	2,839	37.616
Chile[c]	6,264	192	1.625	0.552***	52	2,800	44.645
Lithuania	3,620	114	1.754	0.895*	51	1,992	34.041
Korea, Rep. of	3,322	125	1.942	0.630***	75	2,046	38.286

Belgium (Flemish)[c,e]	4,788	139	2.201	0.945**	39	1,596	34.268
Jordan	3,962	125	2.575	1.041**	72	2,000	49.651
Ghana[c]	2,808	113	5.910	1.691***	24	993	57.266
Hong Kong, SAR	3,477	91	6.412	3.485*	683	1,521	53.729

Source: Author's regressions and calculations based on TIMSS 2003 dataset. See also table 7-1.

*Significant at the 10 percent level; ** significant at the 5 percent level; *** significant at the 1 percent level.

a. Except where otherwise indicated, imputations at the school level were not possible, and thus the regression includes imputed values at the student level only. Fewer than 100 schools were sampled in England and in the U.S. state of Indiana. Inferences from such small school samples seem problematic. Moreover, the regression sample comprises fewer than 100 schools in Hong Kong and the Netherlands.

The following countries (or their states, provinces, or regions) are OECD members: England, the U.S. state of Indiana, the Netherlands, the Repuplic of Korea, and Belgium.

b. Coefficents multiplied by 100.

c. Regression includes imputed values both at the student and at the school level.

d. In the U.S. state of Indiana, no information was provided on the experience of the teachers. The regression therefore does not control for this variable.

e. The regression for Belgium does not control for teacher education since there is no variation in this variable in Belgium.

Is the Relationship Different for Different Student Populations?

Educational systems are regarded by many people as a means of providing equal opportunity for all, irrespective of socioeconomic or sociocultural backgrounds. This mandate of educational systems is often associated with effecting of distributive justice. However, there are also economic justifications for supporting educational systems. When individuals do not receive the education that corresponds to their abilities, part of their potential human capital is lost, and, therefore, they contribute less to the economic prosperity of the country than they otherwise would. As long as other important goals, such as high average performance, are not adversely affected, equality of educational opportunities is a sensible aim also on the basis of economic reasoning. In countries that experience a lot of immigration, another goal of educational systems is to accelerate the process of integration.

How well a country's educational system succeeds in promoting integration and equal opportunities is therefore of interest to every society. Thus, in the context of analyzing the relationship between school size and student performance, the question is whether this relationship differs between groups of students. If students from disadvantaged family backgrounds perform particularly worse in schools of a certain size range, policymakers and educators should consider measures that might be taken to improve the situation.

In the present analysis, differences in the relationship between school size and test performance are considered along two dimensions: the sociocultural and the socioeconomic backgrounds of students. To capture the sociocultural dimension, the analysis distinguishes between two groups: students who never or rarely speak the language of the test at home (minority-language students) and those who always or mostly speak the language of the test at home (majority-language students). A dummy was constructed in which the value 1 is assigned if the test language is never or rarely spoken; the value 0 is assigned otherwise. Thus the focus is on minority-language students—those whose families originated from countries other than the country of the test.

The socioeconomic background of the students was proxied by the number of books in their home. Here, the analysis distinguishes between students who have between zero and twenty-five books in their home and those who have more books in their home.

Each dimension of family background was considered in a separate analysis while the control variables at the student, teacher, and school level remained mainly as before.[25] In each analysis, a dummy was constructed to depict student membership in one of two groups: one comprising students with an advantageous family background and the other comprising students with a less privileged family background. This dummy was entered into the regression equation as an additional constant to capture average level differences in test performance between the two groups. Moreover, to capture the possible difference in the relationship between school size and student test achievement between students from advantaged and disadvantaged backgrounds, the dummy for group membership was interacted with school size and also with its square. Thus the regression equation allowed the relationship between school size and test performance to differ between the two groups of students.

Differences between Students from Different Sociocultural Backgrounds

For most countries, the hypothesis that the relationship is significantly different between language-minority and language-majority students had to be rejected. However, for ten countries, the conclusion was that although school size and student test performance were entirely unrelated for students that are native or almost native in the language of the test, for the group of language-minority students, a relationship did exist. In two of these ten countries, the relationship between size and performance for the group of language-minority students was rather linear, while in the remaining eight countries, the relationship was best described by a quadratic function. Only in three countries—South Africa, Indonesia, and Macedonia—was a statistically significant quadratic relationship between size and performance found for both groups of students. (See table 7-4.)

In South Africa, the results show that for students who speak the test language always or almost always at home, the relationship between school size and test achievement is inversely U-shaped as before, with an optimal school size at around 1,143. For students who never or rarely speak the language of the test at home, the relationship between school size and test performance is positive over the entire range of school size observed, but the relationship becomes increasingly flatter with greater school size. Taken together, as schools become larger, minority-language students gain relative to their advantaged majority-language peers, and the differences in test performance between the two groups decrease.

In Indonesia the relationship between school size and student performance is inversely U-shaped for both groups of students. However, for minority-language students, the relationship peaks much earlier than for majority-language students. For the second group, the relationship is positive for almost the entire range of school size observed and only becomes negative for very large schools. Interestingly, the minority-language group performs better on average than the group of native speakers. However, this pattern reverses for large schools, where minority-language students lose against native speakers, with lower performance as schools become very large.

Table 7-4. *Shape of Relationship between School Size and Student Performance*[a]

| | | Differences between groups of students | | | |
| | | Sociocultural | | Socioeconomic | |
Country or region[b]	*Overall*	*Majority-language students*	*Minority-language students*	*High-SES students*	*Low-SES students*
Armenia	..	..	..	..	..
Australia	Inversely U-shaped	..	CPR	..	DPR
Bahrain	CPR	..	..	..	..
Belgium (Flemish)[c,d]	..	..	..	..	Inversely U-shaped
Botswana[c]	..	..	DNR	..	DNR
Bulgaria	CPR	..	..	..	..
Chile[c]	..	..	..	..	..
Chinese Taipei	..	..	..	..	..
Cyprus	..	..	..	..	..
Egypt	CNR	..	Inversely U-shaped	..	..
England	CPR	..	..	..	..
Estonia[c]	CPR	..	..	..	U-shaped
Ghana[c]	CPR	..	..	DPR	DPR
Hong Kong, SAR	..	..	..	..	..
Hungary	..	..	..	..	CNR
Indonesia[c]	Inversely U-shaped	Inversely U-shaped	Inversely U-shaped	..	..
Iran, Islamic Republic of[c]	..	..	..	..	..
Israel	..	..	U-shaped	..	..
Italy	..	..	Inversely U-shaped	..	..
Japan[c]	..	..	..	..	U-shaped
Jordan	CPR	..	..	..	..

Korea, Rep. of	CPR	…	U-shaped	…	…
Latvia	Inversely U-shaped	…	…	Inversely U-shaped	Inversely U-shaped
Lebanon[c,e]	CPR	…	…	…	…
Lithuania	CNR	…	…	…	…
Macedonia, Rep. of	CPR	INR (almost CNR)	Inversely U-shaped	…	…
Malaysia	…	…	…	…	…
Modova, Rep. of	…	…	…	…	…
Morocco[c]	CPR	…	…	…	…
Netherlands	…	CPR	…	…	…
New Zealand	…	…	CPR	…	CPR
Norway[c,f]	…	…	…	…	…
Palestinian National Authority[c]	…	…	…	…	…
Philippines	…	…	…	…	Inversely U-shaped
Romania	…	…	…	…	…
Russian Federation[c]	…	…	Inversely U-shaped	Inversely U-shaped	…
Saudi Arabia	…	…	…	…	U-shaped
Scotland	…	…	…	…	…
Serbia[c]	…	…	…	…	…
Singapore[c]	U-shaped	…	…	…	CPR
Slovak Republic	…	…	…	…	…
Slovenia[c]	…	…	U-shaped	…	…
South Africa[c]	Inversely U-shaped	Inversely U-shaped	DPR	…	…
Sweden	…	…	…	…	…
Syrian Arab Republic[c]	…	…	…	…	…
Tunisia[c]	CPR	…	…	…	…
United States	IPR	…	…	…	…

(continued)

Table 7-4. *Shape of Relationship between School Size and Student Performance*[a] *(Continued)*

| | | Differences between groups of students | | | |
| | | Sociocultural | | Socioeconomic | |
Country or region[b]	Overall	Majority-language students	Minority-language students	High-SES students	Low-SES students
Benchmark regions					
Basque Country, Spain	Inversely U-shaped	...	...	...	...
Indiana, United States[c,g]	CNR	...	...	...	...
Ontario, Canada	...	...	Inversely U-shaped	...	...
Quebec, Canada	...	...	...	U-shaped	U-shaped

Source: See table 7-1.

a. Except where otherwise indicated, imputations at the school level were not possible, and thus the regression includes imputed values at the student level only. Fewer than 100 schools were sampled in Bahrain, Cyprus, England, Morocco, Scotland, and the U.S. state of Indiana. Inferences from such small school samples seem problematic. Moreover, the regression sample comprises fewer than 100 schools in Armenia, Hong Kong, Moldova, the Netherlands, and the Basque Country in Spain. Abbreviations are used to illustrate more plastically the results from the analysis; in this context, the word "return" does not imply the existence of a causal relationship. CPR, constant positive returns to size; relationship is positive and linear over the entire range of school size observed. CNR, constant negative returns to size; relationship is negative and linear over the entire range of school size observed. DPR, decreasing positive returns to size; relationship is positive over the entire range of school size observed, but the slope of the relationship becomes flatter as schools become larger. DNR, decreasing negative returns to size; relationship is negative over the entire range of school size observed, but the slope of the relationship becomes flatter as schools become larger. IPR, increasing positive returns to size; relationship is positive over the entire range of school size observed, but the slope of the relationship becomes steeper as schools become larger. INR, increasing negative returns to size; relationship is negative over the entire range of school size observed, but the slope of the relationship becomes steeper as schools become larger.

b. The following countries or their states, provinces, or regions are OECD members: Australia, Belgium, Canada, England, Hungary, Italy, Japan, Republic of Korea, the Netherlands, New Zealand, Norway, Scotland, Slovak Republic, Spain, Sweden, and the United States.

c. Regression includes imputed values both at the student and at the school level.

d. The regression for Belgium does not control for teacher education since there is no variation in this variable in Belgium.

e. In Lebanon no information was provided on the education of the teachers. The regression therefore does not control for this variable.

f. In Norway no information was provided on the percentage of students from disadvantaged family backgrounds. The regression therefore does not control for this variable.

g. In Indiana no information was provided on the experience of the teachers. The regression therefore does not control for this variable.

In Macedonia the relationship between school size and student performance is inversely U-shaped for minority-language students. The calculated optimal school size for this group lies at an enrollment level of 1,120 students. Thus minority-language students on average perform best at middle-sized schools, with test scores decreasing as schools become either smaller or larger. For students who are native or almost native speakers in the language of the test, the relationship between size and performance is negative over the entire range of school size observed, with the loss in test performance slightly increasing as schools become larger. However, the curve is only very mildly bent. This indicates that majority-language students in Macedonia perform best in small schools.

In ten other countries, the relationship between school size and student performance could be significantly estimated only for the minority-language group. Thus, in these countries, evidence suggests that school size is unrelated to student achievement for the group of students who are native or almost native speakers. In contrast, for minority-language students, test performance seems to be connected with school size.

A U-shaped relationship between school size and performance for the minority-language group is found in Israel, Latvia, and Slovenia. In these three countries, minority-language students perform on average better in smaller or larger schools, while their performance is lowest in middle-sized schools.

In contrast, an inversely U-shaped relationship for the minority-language students is found in the Canadian province of Ontario and in Italy, Russia, and Egypt. In those locations, students who seldom or never speak the test language at home perform on average better in middle-sized schools while their performance is weaker in smaller or larger schools. However, in the Russian Federation, the minority-language students' test performance initially increases as schools become larger, but the trend reverses at relatively low levels of enrollment. Thus, for most of the range of school size observed in Russia, the relationship between school size and test performance is negative for minority-language students. In Egypt, the peak of the relationship between size and performance is at relatively high levels of enrollment. This means that over a wide range of school size observed, minority-language students on average perform better as schools become larger. Only at very high levels of enrollment does the relationship become negative.

A decreasing negative relationship was found in Bulgaria. There, the relationship between school size and student achievement is negative over the whole range of school sizes observed, but at the same time, the link becomes weaker, meaning that for high levels of enrollment, the loss in test score points in the minority-language group decreases as schools become very large.

Although the quadratic relationship for minority-language students could be significantly estimated for New Zealand, the pattern of the relationship between school size and student performance can more easily be described as positive and linear. The relationship is positive over the whole range of school sizes observed, and the slope of the function describing the relationship is almost constant, that is, almost a straight line. It is therefore concluded that on average the performance of minority-language students increases as schools become larger.

For the countries where a quadratic relationship to describe the differences between the two groups of students could not be fitted to the data, a linear specification was estimated. The results were significant only in Australia. Also in this case, the existence of a relationship between school size and student performance is supported only for the minority-language group. In Australia the relationship between school size and student performance is positive for students from minority-language backgrounds, that is, test performance increases as schools become larger. Language-minority students gain on average 5.7 test score points as school size increases by 100. For majority-language students, however, the school size does not seem to play a role.

The results for the minority-language students indicate that school size is an issue when it comes to equality of educational opportunities. In thirteen countries, a relationship between school size and performance existed for this group while only in three cases was a relationship also found for the majority-language group. Closer investigation is needed to analyze why the differences in the relationship between school size and student performance exist between the groups and how the situation for the disadvantaged students could be improved.

Differences between Students from Different Socioeconomic Backgrounds

The second question relating to equality of educational opportunity is whether the relationship between school size and student performance differs for students from different socioeconomic backgrounds. The socioeconomic status (SES) of the students' families was proxied by the number of books in their homes.[26] The analysis distinguishes between two groups of students: those having more than twenty-five books in their homes (students from high-SES families) and those from homes with fewer books (students from low-SES families).

The relationship between school size and student performance could be described as a quadratic function for both high- and low-SES students only in four places—Lebanon, Saudi Arabia, Ghana, and the Canadian province of Quebec. (See table 7-4.) In Lebanon the relationship between school size and student performance could be described as inversely U-shaped for both groups of students. Optimal school size is estimated at 1,811 for low-SES students and at 2,180 for high-SES students. The difference between the two groups is highest in middle-sized schools and decreases as schools become smaller or larger.

In Saudi Arabia, the relationship patterns for the two groups are opposites. The relationship between school size and student performance is inversely U-shaped for high-SES and U-shaped for low-SES students. Since the optimal and the nonoptimal school sizes for the two groups are very close—438 for high-SES and 328 for low-SES students—the difference in average performance between the two groups is highest in middle-sized schools. High-SES children perform best in middle-sized schools whereas low-SES children perform lowest in schools of this size.

In Ghana the relationship between school size and performance for both low- and high-SES students is inversely U-shaped. However, the optimal school size of 835 for high-SES children is already located at the far right of the school size distribution in Ghana. Therefore the relationship for this group also can be summarized as positive over the whole range of school size observed, with diminishing returns toward the upper end of the size distribution. The same applies to the relationship between size and performance observed for the group of low-SES students. On average, students from both high- and low-SES backgrounds perform better in larger schools.

The pattern of the relationship is U-shaped for both groups in Quebec. However, for high-SES students, average performance decreases only slightly as school size approaches an enrollment of 500; thereafter the relationship becomes positive. In contrast, the decrease in average performance is significant for low-SES children as school size increases from small to middle sized. The test scores of children from families with a low socioeconomic background are lowest in middle-sized schools with an enrollment of around 1,194 students. Thus for high-SES students size is only weakly connected with test scores at schools with small to middle-sized enrollment levels while low-SES students perform much better in small or large schools.

In six countries—Botswana, the Philippines, Japan, Estonia, Bulgaria, and Australia—the quadratic relationship between school size and student achievement was only significant for children from low socioeconomic backgrounds. For children from better-off families, test performance was entirely unrelated to school size in these countries.

In Botswana and in the Philippines, the relationship between school size and student achievement was inversely U-shaped for children from low-SES families. In both countries, children from low-SES families perform on average best in middle-sized schools. On the other hand, in Japan and Estonia, the relationship between size and performance was U-shaped for children from low-SES families. This implies that average performance for this group of students is highest in small and large schools. A quadratic relationship could also be estimated for low-SES students in Bulgaria. However, the relationship is negative over the whole range of school size observed. On average, children from this group perform worse as schools grow larger, and performance is lowest in the largest schools. At the same time, however, the decline in test performance observed with increasing size diminishes as schools become larger. In Australia, the relationship for low-SES students is positive over the whole range of school size observed but diminishing in absolute terms. On average, student performance in this group increases as schools grow larger and is highest in the largest schools.

In three countries, the relationship between school size and performance for the group of low-SES students was better described using a linear form. While for children from better-off families such a relationship does not seem to exist, a significant relationship for children from low socioeconomic backgrounds is found in Singapore, New Zealand, and Hungary. The linear relationship is positive in Singapore and New Zealand, implying that children from low-SES families perform better on average as schools become bigger. In Hungary, however, the estimated relationship is negative. There, children from low-SES families perform better in smaller settings.

Conclusion

Much of the existing research on the effects of school size on student performance has been confined to the United States, where evidence suggests that students perform best at schools with an enrollment between 600 and 900. This raises the question of whether the experience of other countries also provides evidence for the existence of an optimal school size range. Thus the present analysis examined the relationship between school size and student performance on an international level, using data from countries that participated in TIMSS 2003. Table 7-4 summarizes the findings discussed.

Because school size can be hypothesized to correlate with other factors that also affect student performance, several control variables were included that accounted for the family background of the student and for teacher and school effects. Only in six out of fifty-one participating countries and benchmark regions could a quadratic relationship between size and performance be estimated. A linear relationship could be significantly estimated for fourteen countries. Thus this study did not find evidence of a systematic relationship between school size and student achievement that is stable across countries. Rather, if school size is related to student achievement, this relationship is different across countries.

A further question addressed here was whether the relationship between school size and student performance varied between groups with different sociocultural and socioeconomic backgrounds. A dummy for group membership (advantageous background versus disadvantageous background) was constructed in each case and interacted with school size and school size squared. Again, no clear pattern emerges that applies to all countries. However, one important finding is that the relationship between school size and student achievement is indeed different for groups of students that vary with respect to sociocultural or socioeconomic status. This result should be examined more closely on the individual country level to develop strategies to improve the performance of students from disadvantaged family backgrounds.

An interesting issue that could not be addressed within this study is the question why schools of a certain size might perform better than others. This question remains open for now, awaiting further exploration using information on the extent of autonomy within schools and on the student and teacher perceptions of the atmosphere within the schools.

Notes

1. The discussion in this paragraph, and in the next, is based on Kathleen Cotton, "School Size, School Climate, and Student Performance," School Improvement Research Series 20 (Portland, Ore.: Northwest Regional Educational Laboratory, 1996).
2. Note, however, that not all educational systems offer the same possibility or the same degree of ability grouping and curriculum targeting.
3. Valerie E. Lee and Julia B. Smith, "High School Size: Which Works Best and for Whom?" *Educational Evaluation and Policy Analysis* 19 (Autumn 1997): 205–27.
4. It should be noted that the institutional settings and the degree of school autonomy not only vary between countries but also within them.
5. See the review by Matthew Andrews, William Duncombe, and John Yinger, "Revisiting Economies of Size in American Education: Are We Any Closer to a Consensus?" *Economics of Education Review* 21, no. 3 (2002): 245–62.
6. See, for example, Lee and Smith, "High School Size."
7. The development of the test contents was a cooperative process involving national research coordinators from all participating countries, and the curriculum framework was endorsed by all participating countries. The study also performed a test-curriculum matching analysis that restricted the analysis to items definitely covered in each country's curriculum; the results indicated little difference for the overall achievement patterns. For more detailed information, see Ina V. S. Mullis and others, *TIMSS 2003 International Mathematics Report: Findings from IEA's Trends in International Mathematics and Science Study at the Fourth and Eighth Grades* (Boston College, 2004), p. 180.
8. However, there are two exceptions in the TIMSS 2003 for the eighth grade: for Syria the sample characteristics are not completely known while England did not meet the school participation rate. See Mullis and others, *TIMSS 2003 International Mathematics Report,* pp. 31–32.
9. Pierre Foy and Marc Joncas, "TIMSS 2003 Sampling Design," in *TIMSS 2003 Technical Report,* edited by Michael O. Martin, Ina V. S. Mullis, and Steven J. Chrostowski (TIMSS and PIRLS International Study Center, Boston College, 2004), pp. 109–22.
10. The benchmarking participants for the eighth grade include the Basque Country in Spain, the U.S. state of Indiana, and the provinces of Ontario and Quebec in Canada. For the fourth grade, the benchmarking participants were again Indiana in the United States and Ontario

and Quebec in Canada. With the exception of Yemen, all countries that participated at the fourth-grade level also participated at the eighth-grade level. England and Scotland are treated as individual countries since they have separate school systems that participated separately in the tests. In Belgium only the Flemish school system participated in TIMSS 2003.

11. For an introduction to item response theory, see, for example, Wim J. van der Linden and Ronald K. Hambleton, eds., *Handbook of Modern Item Response Theory* (New York: Springer, 1997).

12. See Eugenio J. Gonzalez, Joseph Galia, and Isaac Li, "Scaling Methods and Procedures for the TIMSS 2003 Mathematics and Science Scales," in *TIMSS 2003 Technical Report,* edited by Martin, Mullis, and Chrostowski, pp. 252–73.

13. Michael O. Martin, ed., *TIMSS 2003 User Guide for the International Database* (International Association for the Evaluation of Educational Achievement and TIMSS and PIRLS International Study Center, Boston College, 2003).

14. Without these countries, the average number of schools sampled in each country is 151.

15. For one school in Morocco and one in the Syrian Arab Republic, the value for school size was incorrectly coded as 9999 instead of being coded as "missing." These were accordingly recoded as missing. Obvious outliers in the distribution of school size were discarded.

16. Foy and Joncas, "TIMSS 2003 Sampling Design."

17. For this problem of hierarchical data structure see, for example, Brent R. Moulton, "Random Group Effects and the Precision of Regression Estimates," *Journal of Econometrics* 32, no. 3 (1986): 385–97.

18. See Halbert White, *Asymptotic Theory for Econometricians* (Orlando, Fla.: Academic Press, 1984). Cited in Angus Deaton, *The Analysis of Household Surveys: A Microeconometric Approach to Development Policy* (Johns Hopkins University Press, 1997).

19. Foy and Joncas, "TIMSS 2003 Sampling Design."

20. See William H. DuMouchel and Greg J. Duncan, "Using Sample Survey Weights in Multiple Regression Analyses of Stratified Samples," *Journal of the American Statistical Association* 78, no. 383 (1983): 535–43. Cited in Jeffrey M. Wooldridge, "Asymptotic Properties of Weighted *M*-Estimators for Standard Stratified Samples," *Econometric Theory* 17, no. 2 (2001): 451–70.

21. Only the highest education level of either the mother or the father was considered in the analysis.

22. The same regressions were also estimated without using imputed values. In most cases, the results were not different from those reported here except for the standard errors, which were larger in the sample without imputed values.

23. Lee and Smith, "High School Size."

24. However, the number of PSUs is relatively low in Bahrain and the Basque Country in Spain, with only 50 and 87 schools in the regression sample, respectively.

25. In the regressions that investigate differences between groups along the sociocultural dimension, the dummies for whether the parents and the students were born in the country of the test were excluded. In the regressions investigating differences between groups along the socioeconomic dimension, the dummies for the books at home categories and for parental education were excluded.

26. For a more detailed discussion on the use of the number of books at home as a proxy for family background, see Gabriela Schütz, Heinrich W. Ursprung, and Ludger Wößmann, "Education Policy and Equality of Opportunity," Working Paper 1518 (Munich: CESifo, 2005), and the references listed therein.

8

Examining Educational Technology and Achievement through Latent Variable Modeling

ELENA C. PAPANASTASIOU AND EFI PAPARISTODEMOU

The use of computers in the teaching and learning of mathematics appears to be increasing exponentially year by year. Consequently, the amount of research on the relationship between computer use and achievement has been exponentially increasing as well. This trend is reflected by the change in how the International Association for the Evaluation of Educational Achievement (IEA) has approached the computer-achievement relationship. In 1984 when the Second International Science Study was conducted, a single question was included regarding the availability of computers in schools.[1] By 1995 the student background questionnaire from the Third International Mathematics and Science Study (TIMSS) included four very broad computer-related questions. Two of the questions asked about the frequency with which the students used computers in their mathematics and science classes. The other two questions asked whether students liked using computers for mathematics and for science. In 1999 the student background questionnaire of the Third International Mathematics and Science Study–Repeat (TIMSS-R) included three times as many computer-related questions, which were much more specific compared to the TIMSS-1995 questionnaire. In the 2003 TIMSS, the number of computer-related questions did not increase, but they became even more specific, focusing on the locations in which computers are used by students, as well as on the types of activities that are being performed on the computer, with an emphasis on the use of the Internet.

The inclusion of such questions reflects the increasing and more detailed attempts to examine how computer environments could be effectively used to improve students' understanding of mathematics and science. However, this is not a simple task since the relationship between computer use and school-related variables such as student achievement is constantly transformed by the ever increasing presence of computers in everyday life, and by society's increasing reliance on computer technologies. As a result, educators must constantly adapt their practices to conform to the changing educational needs of the students.

The purpose of this study is to propose a model concerning the current relationship between technology use and mathematics achievement and test it with data from TIMSS 2003. This model is examined in four countries: the United States, Cyprus, the Russian Federation, and South Africa. The present study attempts to enhance the current understanding of how the technology-achievement relationship has evolved in recent years, during which the number of computers at home and in school settings has increased, and technology has become an integral part of daily life.

Educational Technology and Mathematics

The key feature of a computer-based environment is that it presents a formal, computable representation of mathematical objects and relationships. Papert's early work has been very influential in the development of computer environments called microworlds.[2] According to Papert, learners interact with the microworld and build their own computer-based models. These models reflect the learners' thinking about the mathematical objects and relationships as they work on particular activities.

Noss and Hoyles define a computer environment as a flexible, interactive, expressive medium for working with mathematical objects and operations.[3] Computer-based environments provide access to formal mathematical knowledge through the nature of the "intermediate" screen objects with which students interact in order to construct and manipulate new objects and relationships. Moreover, these environments allow the learner to explore simultaneously the structure of the accessible objects, their relationships, and the representations that make them accessible. It can be said that the microworld "evolves" as the learner's knowledge grows.

According to Lajoie, Jacobs, and Lavigne, computer-based learning environments support the "learning by doing" philosophy.[4] For example, in a computational modeling approach to statistics, a modeling language and various sets of associated tools are made available to learners, allowing them to pursue personally meaningful investigations. Learning by doing involves building up mental structures so that concepts may become linked into a mental network that

allows some ideas to be assimilated readily while others become radically transformed by the assimilating structure. If a concept that is taught is well assimilated to the teacher's internal structure, but the learners' structures differ
sufficiently from the teacher's, then what is taught will be radically transformed,
and there will not be an effective link with the student's mental network.[5] Arguing along similar lines, Harel and Papert describe "instructional software" in
which "the communication between the software producers and their medium
is dynamic. It requires constant goal-defining and redefining, planning and
replanning, representing, building and rebuilding, blending, reorganizing, evaluating, modifying, and reflecting in similar senses."[6]

In other words, instructional technology should be used as a computer-based
environment to help students construct ideas. Children who are educated in a
computer-based environment have the opportunity to express their own representations and also to reconstruct and redesign the computer environment without losing the aim of each activity in which they are engaged. This ownership
helps to give meaning to the task and enables children to feel that they own the
ideas developed during their interaction with the task.

Calculator Use and Mathematics

Calculators can be considered as one of the oldest modern technological devices
still being used in mathematics classes. The advent of calculator technology has
influenced the teaching of mathematics in a profound way.[7] In the case of graphic
calculators, previous studies have concluded that it is possible to improve students' understanding of variables by giving them environments in which they
could manipulate examples, predict, test, and gain experiences on which higher-
level abstractions could be built.[8] The graphic calculator's value lies in its intrinsic use of variables in its operations and in its multiline display that allows the
user to see, reflect upon, and react with several sequential inputs or outputs.

Research on the use of calculators with younger students (11- to 12-year-
olds) has also shown promising results. According to Cedillo, the use of graphic
calculators for given algebraic tasks helped children gain an awareness of the gen-
erality of algebraic expressions.[9] The results of the same study found that calculators even helped children gain some understanding of inverse functions and
algebraic equivalence. Elliott, Hudson, and O'Reilly have investigated the use
of graphic calculators to explore their effects on students' visualization skills in
the context of graphing functions. They found that the students developed
positive perceptions of the validity of visual methods in mathematics.[10]

Technology Use and Achievement

Since educators first began to use computers in the classroom, researchers have
tried to evaluate whether computer use has a positive impact on student

achievement.[11] There is plenty of evidence to indicate a positive relationship between technology and student achievement.[12] However, most of these studies emphasize that for technology to have an effect on achievement, it must be challenging and focused on higher-order thinking skills, and the teachers must be capable of using and teaching it and have the appropriate support. In other words, examining computer use or technology by itself is not enough to determine its effects on student achievement. What seems to be important, however, is the way in which technology is used.

In addition to how technology is being used, the conclusions that can be reached about the interrelationship of computer use and achievement are limited by the research methods that are used, as well as by the types of analyses performed on the data. Consequently, although prior studies have found positive correlations between computer use in school and achievement, those correlations do not necessarily imply cause-and-effect relationships since many of those studies were not experimental.[13] Based on the same rationale, nonexperimental studies that found negative relationships between computer use in school and achievement do not imply that computer use decreases student achievement.[14]

For example, the 1995 TIMSS results indicated that in Cyprus, Hong Kong, and the United States, fourth-grade students who used computers most frequently in the classroom were those with the lowest scores in their countries.[15] Although this was a surprising result, it did not leave much room for explanation because there were not enough details on how computers had been used in those classrooms. One possible hypothesis was that teachers assigned the use of computers to low-ability students who were not able to catch up with the instruction offered to the rest of the class. By analyzing the same dataset, Pelgrum and Plomp found that frequent use of the computer was strongly associated with student-centered pedagogical approaches, although it was not clear why such activities were associated with lower achievement.[16]

Another study by the Educational Testing Service reported that students who spent more time on computers in school actually performed slightly worse than those who spent less time on them.[17] This same study found that the eighth graders who used computers primarily for "drill and practice" scored more than half a grade lower than students who did not use them in that way. For fourth-grade students, however, the drill and practice software had a small positive impact on their performance.

Ravitz, Mergendoller, and Rush also wanted to explore whether there is a positive or negative relationship between achievement and student computer use.[18] The results of this study found that there is a negative relationship between in-school computer use and student achievement. However, the authors found a positive overall relationship between student achievement and computer proficiency, as measured by the student's reported ability to use a variety of software. In turn, the ability to use software was related to the frequency of use in both the home and school.

There is no doubt that technology (in the form of either computers or calculators) has the potential to improve students' educational experiences, as well as to improve people's everyday lives. However, educational technology must be used by students and educators appropriately, since it is the way in which technology is used that can have an effect on academic achievement.[19] Educators using technology need to shift the focus from instruction to "construction." This is especially important since some studies have found inverse relationships between school computer use and achievement.[20] Such findings confirm that the positive impact of technology does not come automatically and that it should not be expected to affect all groups and nations equally.[21] For example, the impact of computers on achievement in the Republic of Korea, where 98 percent of the students own computers, could not be directly compared to that of Indonesia, where only 17 percent of students own computers. Nor could one compare the Islamic Republic of Iran, where only 2 percent of students use computers both at home and at school, with Hong Kong, where 89 percent of students use a computer at both locations.

Therefore, when one makes country comparisons, the students' background as well as the characteristics of the educational systems should be taken into account when one interprets the unique relationships between computer use and achievement. In this chapter, the TIMSS 2003 dataset is used to analyze how these relationships have evolved in the United States, Cyprus, South Africa, and the Russian Federation.

Methods

TIMSS 2003 is the third cycle of a continuing series of international mathematics and science assessments conducted every four years. TIMSS assesses achievement in about fifty countries around the world and collects a rich array of information about the educational contexts for learning mathematics and science. This is done by asking students, their teachers, and school principals to provide information through questionnaires about the curriculum, schools, classrooms, and instruction. The aim of TIMSS is "to improve the teaching and learning of mathematics and science by providing data about students in relation to different types of curricula, instructional practices, and school environments."[22]

The mathematics assessment framework of TIMSS is based on two dimensions. The first is the cognitive dimension, defined as the set of behaviors that the students are expected to perform while engaging with the mathematics content. The cognitive dimension includes knowing facts and procedures, using concepts, solving routine problems, and reasoning. The content dimension defines the specific mathematics subjects that are covered in the assessment. It includes the content domains of numbers, algebra, measurement, geometry, and data.

TIMSS used item response theory (IRT) to summarize the achievement results, with a mean score of 500 points and a standard deviation of 100. The IRT method of scaling allowed the performance of students across countries to be summarized on a common metric, although not all students had the same items on the mathematics test. IRT also enabled the accurate measure of trends based on previous assessments. A set of two- and three-parameter models was used for the dichotomously scored multiple choice items, while a generalized partial credit model was used for the construct-response items.[23]

The sample of students used in this study comes from population 2 of TIMSS, which includes eighth-grade students. These students tended to be between 14 and 15 years old.

Country Characteristics

To comprehend the results of the study, it is important to have some background information on the countries whose students are compared; this is presented in table 8-1. Cyprus was included as a representative of a small, average-income country, with the large majority of the students having computers at their homes (82 percent). Despite the prevalence of computers, the average performance of Cypriot students on TIMSS (459) was below the international average (467). South Africa and the Russian Federation are similar in that they are low-income countries, with only about one-third of their students owning computers. However, the students from the Russian Federation obtained an average score of 508 on the TIMSS mathematics test, which was above the international average. The South African students only scored 264 on the TIMSS mathematics test, which was the lowest mathematics achievement level among all countries that took part in TIMSS 2003. Finally, the United States is an example of a high-income country, with almost all of the students in the sample owning computers (93 percent). The average performance of U.S. students on the test was above average, with a mean score of 504. The sample sizes and proportion of female students for each country are shown on table 8-1.

Structural Equation Modeling

Structural equation modeling (SEM) is a statistical technique that can be used in theory development because it enables researchers both to propose and test theories about the interrelationships among variables in multivariate settings.[24] The use of such models is especially helpful since it allows simultaneous exam-

Table 8-1. *Characteristics of Country Samples, 2003*[a]

Characteristics	Cyprus	Russian Federation	South Africa	United States
Units as indicated				
Sample size (no. of students)	4,002	4,667	8,840	8,912
Percent female	48.6	50.0	50.6	51.9
Population (millions)	0.8	144.1	45.3	288.4
GNI per capita (U.S. dollars)	12,320	2,130	2,500	35,400
Average student age (years)	13.8	14.2	15.1	14.2
Percentage of students	82	30	37	93
having computers	(0.6)	(2.0)	(1.3)	(0.4)
Average TIMSS score	459	508	264	504
	(1.7)	(3.7)	(5.5)	(3.3)
Average frequency of student involvement in technology-related activities				
Look up ideas and information for math ($\bar{x}$)	2.76	1.86	3.68	2.13
Write reports ($\bar{x}$)	3.47	2.44	3.12	3.16
Process and analyze data ($\bar{x}$)	3.30	2.22	3.01	2.51
Play computer games ($\bar{x}$)	2.55	2.14	1.81	2.29
Use the Internet ($\bar{x}$)	2.39	1.47	1.87	2.99
Use calculators ($\bar{x}$)	1.80	1.94	2.60	2.95

Sources: Data in the first half of the table are from Ina V. S. Mullis and others, *TIMSS 2003 International Mathematics Report: Findings from IEA's Trends in International Mathematics and Science Study at the Fourth and Eighth Grades* (Boston College, 2004). Values shown in the second half of the table are authors' calculations.

a. Standard errors are shown in parentheses. $\bar{x}$ represents the mean response of the students in each country.

ination of multiple relationships without the concern of inflated alpha estimates. Some additional advantages of SEM include the reduction of measurement error by including multiple indicators per latent variable, the ability to acknowledge and model error terms, and the ability to concurrently test overall models rather than individual coefficients.

For the purpose of this study, the data were analyzed with the structural equation modeling software AMOS 5.0.1.[25] Maximum likelihood estimation was the procedure used for estimation of the model.

Hypothesized Model

As a first step in this analysis, a model was developed to attempt to explain the data. The conceptual model of this study (see figure 8-1) and the relationships

Figure 8-1. *Hypothesized Model for the Relationship between Technology Use and Mathematics Achievement*[a]

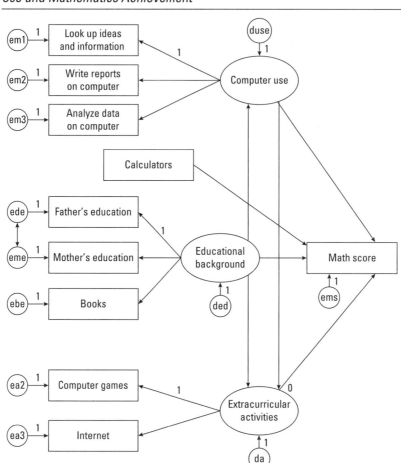

Source: Authors' construction (see text).
a. Maximum likelihood estimation procedure used. Observed variables are shown in rectangles; unobserved and latent variables are shown in ovals. Small circles with the letter "d" that have direct paths to the latent variables are disturbances. The disturbances represent the variation of each latent variable that cannot be explained by the model. Arrows labeled with names such as "em1" and "ede" that have direct paths to the endogenous observed variables represent the errors in the measurement of each of those observed variables.

between the latent variables shown in this model are based on a review of the literature. The outcome of the study is the student's mathematics score, as measured by TIMSS 2003. Based on the model, there are two technology-related latent variables and one technology-related observed variable that are hypothesized to predict the student's mathematics score. The first latent variable is that of computer use, which comprises three items measuring the frequency with which several activities occurred with the use of a computer: "I look up ideas and information for mathematics," "I write reports for school," and "I process and analyze data."

The second technology-related latent variable hypothesized to influence the student's mathematics score was computer-related extracurricular activities. This latent variable comprised two questions: "I play computer games (outside of school)" and "I use the Internet (outside of school)."

The third technology-related variable was the observed variable of using calculators. This was a single question that asked the students the frequency with which they used calculators in their mathematics classes. This variable was purposely excluded from the latent variable of "computer use" so that calculator use could be examined independently of computer use.

Another, nontechnology-related latent variable was included in this model, that of the student's "educational background." This variable comprised three items: the education of the father, the education of the mother, and the number of books in the student's home. Although this variable is not directly related to technology, it was purposely included in the model to control for such variation in the student's background. Socioeconomic status (SES) has been universally acknowledged to affect student achievement in school. However, for the purpose of this study, the educational background, which is one of the indicators included in estimating the SES, was considered more interesting than the SES itself.

Since SEM creates the latent variables, they also must be defined by a scale. This is done by assigning the value of 1.0 to the path from a latent variable to one of its observed variables.[26] Thus, to set the scale for the latent variable of computer use, the path from this variable to the question "I look up ideas and information for mathematics" was set to a value of 1. The question that set the scale for the latent variable of extracurricular activities was that about playing computer games. Finally, the question that set the scale for the latent variable of educational background was that of the father's education.

All observed variables are represented in the structural model with rectangles, while all unobserved and latent variables are represented with ovals. The smaller circles that have direct paths to the latent variables are "disturbances," which represent the variation of each variable that cannot be explained by the model. Finally, the arrows labeled with names such as "em1" and "ede" that have direct paths to the endogenous observed variables represent the errors in the measurement of each of those observed variables. These error terms are very important in the model since no assumptions are made that the variables used in this study are perfectly reliable.

Results

Table 8-1 describes the extent to which students in the four study countries engage in technology-related activities. The students in Cyprus were the ones who responded most frequently that they used computers for writing reports for school ($\bar{x}$ = 3.47), and analyzing data ($\bar{x}$ = 3.30). Cypriot students also played computer games most frequently ($\bar{x}$ = 2.55) in comparison to students from the other countries. However, students in the United States used the Internet most frequently ($\bar{x}$ = 2.99), while South African students looked up information for math ($\bar{x}$ = 3.68) most frequently. South Africa was also the country where students were least likely to play computer games ($\bar{x}$ = 1.81), while it was the Russian students who used the Internet less frequently ($\bar{x}$ = 1.47).

In terms of calculator use, Cyprus was the country where students indicated that they used calculators for mathematics to the least extent ($\bar{x}$ = 1.80, SD = 0.86), while the students in the United States indicated that they used calculators to the largest extent ($\bar{x}$ = 2.95, SD = 1.00). The students in South Africa also used calculators to a larger extent ($\bar{x}$ = 2.60, SD = 0.83), while in the Russian Federation, their use was less frequent ($\bar{x}$ = 1.94, SD = 0.88).

The first step in evaluating the results of the SEM analyses is the examination of the fit criteria, to determine whether the hypothesized model fits each country's data. Six fit indexes were taken into consideration, as well as the percentage of variance (R^2) in mathematics achievement. With the exception of the chi square (χ^2) and the χ^2/d.f. ratio, the rest of the indexes fit the data quite well for all countries. More specifically, the chi-square statistic was significant in all countries, indicating that the data were significantly different from the model. This result was expected, however, since this statistic and its statistical significance are influenced by large sample sizes. The χ^2/d.f. ratio, which is influenced by large samples to a lesser extent, only indicated a good fit for the Russian Federation since its value of 2.13 was less than 2.5.

The root mean square error of approximation (RMSEA) had a good fit for all countries since, according to Hu and Bentler, its values were equal to or less than 0.06.[27] For the normed fit index (NFI), the comparative fit index (CFI), and the Tucker-Lewis index (TLI), values between 0.90 and 0.94 indicate a good fit while values equal to or greater than 0.95 indicate a very good fit. In the case of Cyprus, the NFI and the CFI indicated a good fit while the TLI equaled 0.89. All three indexes indicated a very good fit for the Russian Federation since they were all higher than 0.95 (NFI = 0.99, CFI = 0.96, TLI = 0.99). In South Africa the TLI did not support the fit of the model (TLI = 0.81), although the NFI and the CFI both equaled 0.90, indicating a good fit. Finally, the NFI (0.95) and CFI (0.96) for the United States indicated a very good fit while the TLI (0.92) indicated a good fit. Overall, in all four countries, the

Table 8-2. *Fit Indexes across Countries*

	Cyprus	Russian Federation	South Africa	United States
Units as indicated				
$\chi^{2\,a}$	338.87	61.84	839.92	553.63
χ^2/df ratio	11.69	2.13**	28.96	19.09
NFI	0.94*	0.99**	0.90*	0.95**
CFI	0.94*	0.96**	0.90*	0.96**
TLI	0.89*	0.99**	0.81	0.92*
RMSEA	0.05**	0.02**	0.06**	0.05**
$R^2_{\text{math score}}$ (percent)	38	25	50	43

Source: Authors' calculations.
*Good fit; **very good fit.
a. $p = 0.00$ for all values of χ^2.

majority of the fit indexes supported the fit of the model (when one ignores the χ^2, which is greatly influenced by the sample size). (See table 8-2.)

The percentage of variance of mathematics education explained ($R^2_{\text{math score}}$) ranged from 25 to 50 percent. The least amount of variance explained was in the case of the Russian Federation (25 percent) whereas in South Africa, 50 percent of the variance of the dependent variable was explained. These percentages are quite significant, indicating that the model overall does a good job in explaining a medium to large portion of the variance of the students' mathematics scores on TIMSS.

Since the fit of the model in all four countries has been established, the parameter estimates can then be examined and interpreted. The standardized parameter estimates of the model in the four countries are presented in figures 8-2 through 8-5. These parameters indicate the contribution of each of the latent and observed variables to the overall model. More specifically, they represent the amount by which each endogenous (dependent) variable would change for every standardized unit of change of an exogenous (independent) variable.

The standardized regression weights for all four countries are presented in table 8-3. The path coefficients (β) from the three latent variables (computer use, extracurricular activities, and educational background) to their corresponding observed variables are all adequate, across countries. This indicates that the observed variables are representative of the latent domains they are supposed to measure. These results are presented at the bottom of table 8-3.

The most interesting aspects of the path coefficients, however, are the paths that directly lead to the student's mathematics score. Table 8-3 and figures 8-2

Figure 8-2. *Cyprus Model*

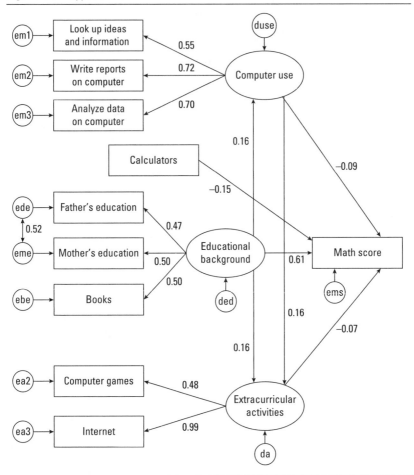

Source: Authors' calculations.
a. See figure 8-1 for explanation of diagram components. Values shown are standardized path coefficients.

through 8-5 demonstrate that the factor that has the greatest effect on the mathematics scores is that of the student's educational background. The highest effect of educational background was found in the United States (β = 0.65), followed by Cyprus (β = 0.61). The lowest effect of educational background was found in the Russian Federation, although its effect was still very high (β = 0.53).

Computer use did not have an effect on the student's math scores in Cyprus or the Russian Federation, although it had a slightly negative effect in South Africa (β = −0.19) and the United States (β = −0.13). Similar results were found

Figure 8-3. *Russian Federation Model*

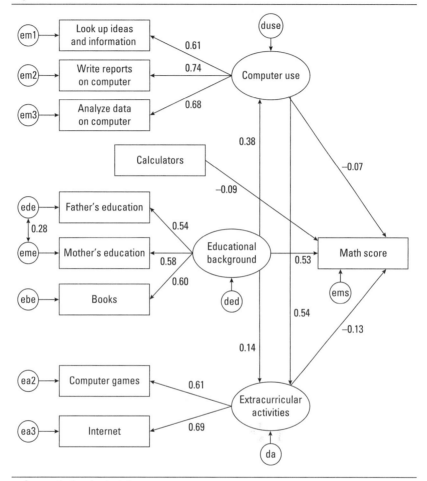

Source: Authors' calculations.
a. See figure 8-1 for explanation of diagram components. Values shown are standardized path coefficients.

with the students' technology-related extracurricular activities. In all four countries, students who spent more time on computer-related extracurricular activities tended to have lower mathematics scores on TIMSS. The largest effect was found in South Africa ($\beta = -0.25$). Finally, calculator use in mathematics was associated with lower mathematics scores in Cyprus ($\beta = -0.15$) and with higher scores in the United States ($\beta = 0.10$). Overall, however, one can see that frequent technology use was associated with lower mathematics scores across countries. The only exception was the United States, where frequent calculator use was associated with higher mathematics scores.

Figure 8-4. *South African Model*

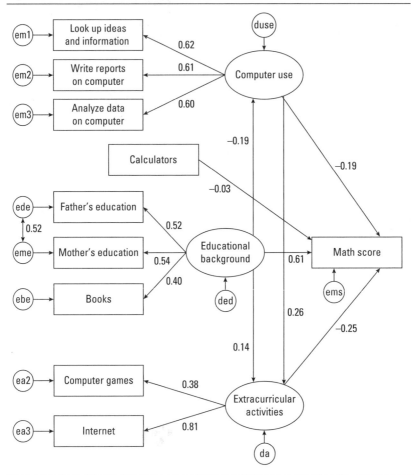

Source: Authors' calculations.
a. See figure 8-1 for explanation of diagram components. Values shown are standardized path coefficients.

Relationships between the three latent variables were also found in this study. Across all countries, students who used the computer frequently for schoolwork were more likely to also use the computer for extracurricular activities. The strongest path existed in the Russian Federation ($\beta = 0.54$), while the path for the United States was very weak ($\beta = 0.09$). Another result of the study was that students from more educated backgrounds were more likely to be involved in technology-related extracurricular activities. A final result of the study was that students from more educated backgrounds were more likely to use computers for their schoolwork. The strongest relationship was found in the Russian Federation

Figure 8-5. *United States Model*

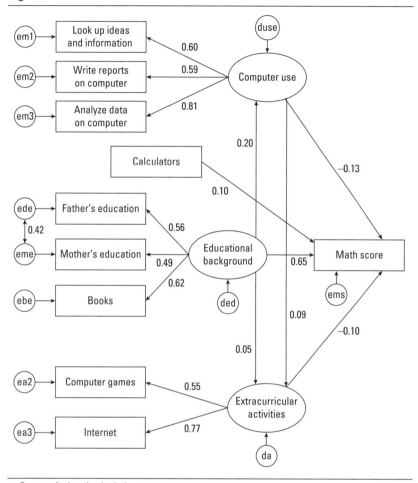

Source: Authors' calculations.

a. See figure 8-1 for explanation of diagram components. Values shown are standardized path coefficients.

($\beta = 0.38$). The only exception was South Africa, where students from more educated backgrounds tended to use the computer less for schoolwork ($\beta = -0.20$).

Discussion

The overall results of the study are not especially encouraging with regard to the overall relationship between technology use and achievement. After controlling for the effects of educational background, the paths between technology use and achievement tend to be small and negative. This is consistent for all countries,

Table 8-3. *Standardized Path Coefficients by Country*

Variable relationships			Standardized path coefficients			
			Cyprus	Russian Federation	South Africa	United States
Computer use	→	Math score	-0.086	-0.065	-0.189	-0.133
Extracurricular activities	→	Math score	-0.074	-0.132	-0.248	-0.100
Educational background	→	Math score	0.614	0.533	0.605	0.654
Use calculators	→	Math score	-0.152	-0.089	-0.026	0.105
Educational background	→	Computer use	0.156	0.377	-0.195	0.200
Educational background	→	Extracurricular activities	0.162	0.144	0.144	0.051
Computer use	→	Extracurricular activities	0.156	0.542	0.262	0.093
Computer use	→	Look up ideas and information	0.553	0.612	0.624	0.600
Computer use	→	Write reports	0.717	0.736	0.609	0.595
Computer use	→	Process and analyze data	0.695	0.684	0.602	0.814
Extracurricular activities	→	Use internet	0.990	0.689	0.808	0.771
Extracurricular activities	→	Play computer games	0.482	0.614	0.382	0.554
Educational background	→	Mother's education	0.503	0.577	0.543	0.494
Educational background	→	Father's education	0.467	0.544	0.517	0.558
Educational background	→	Books at home	0.498	0.604	0.396	0.624

Source: Authors' calculations.

with the exception of one path in the U.S. model, which will be discussed further on. Although these paths do not imply cause-and-effect relationships, they should be examined more closely.

More specifically, the results of this study fall along two lines. First, educational background appears to have a much larger effect on mathematics achievement than any of the technology-related variables. This is not surprising since educational background has always been shown to affect students' school performance in many subjects, as well as their patterns of computer use.[28] Such results, though, always tend to come as a blow to educational systems, which appear incapable of overcoming the effects of SES and educational background. Despite the many years of studies by numerous researchers around the world, the effects of school education on students is not as large as it could be. Based on this result, children of better-educated parents and children from stronger educational backgrounds are advantaged in school while children from less educated backgrounds are disadvantaged. This is quite disappointing, and it makes one wonder how this problem in the field of education can be solved.

The newest trend and hope for education has been that of technology. Technology holds immense potential for improving student performance as well as in improving people's everyday lives. However, to be effective, educational technology has to be used appropriately. Most likely, the full potential of technology in education has not been reached yet. According to the second set of results of this study, the students who use technology (for school work or for leisure) to a smaller extent tend to have higher mathematics achievement on TIMSS than the students who use such technologies more frequently. Although this effect is small, it does exist. Similar results have also been found in previous studies using TIMSS, Program for International Student Assessment (PISA) and other datasets.[29]

The inverse relationship between technology use and student achievement found in this study is in contrast to other studies that tend to praise the use of technology and showcase its positive results.[30] One should keep in mind, though, that although technology might work exceptionally well when it is used by trained educators, it might not work as well in situations where teachers do not adopt the requisite didactical ideas for its application. Sutherland and Balacheff raise the issue of didactical complexity that can arise in computational environments for the learning of mathematics.[31] They argue that computer-based environments can provide access to mathematical worlds as long as the teacher is aware of the types of constructive interactions that take place between the student and the computer environment. When higher-order constructive interactions do not occur, there is no guarantee that the students will be able to benefit from the computer-related activities or computer microworlds.

The negative results of the current study can be explained as follows. On one hand, it is not surprising that frequent student involvement in technology-related extracurricular activities is associated with lower levels of mathematics achievement. Those students who spend large amounts of time using the Internet and playing computer games end up spending less time studying, which therefore leads to lower achievement. This result should not come as a surprise to anyone. However, the surprising result is that the students who used the computer frequently to analyze data, write reports, and look up information also had lower results than the students who performed those activities less frequently. A possible interpretation of this finding is that the specific activities that are performed are not advantageous for the student's learning in mathematics. It is possible that such activities might help increase the student's technological literacy and other related skills, but they are not necessarily helpful to learning mathematics. This is in accord with Wenglinsky, who found that the kinds of thinking skills associated with various activities influence whether or not the students would benefit from technology use.[32] According to Wenglinsky, computer activities that required the use of higher-order thinking skills were more beneficial to the students who used them.

The results of this study reinforce the importance of the proper use of technology. Technology alone does not automatically improve mathematical learning. In particular, a deeper understanding of the ways in which learners structure their own learning is essential for the effective computer environments. If this understanding shapes the instructional use of technology, then learners can draw upon and reconstruct the material in ways that they consider appropriate in order to derive meaning for the mathematical subject areas that they are being taught.[33]

The use of calculators in mathematics did not have any practically significant effect in the Russian Federation and in South Africa. However, although the use of calculators was negatively associated with mathematics achievement in Cyprus, it was positively associated with achievement in the United States. The educational system in Cyprus views calculators as an impediment to student learning in mathematics. Therefore, although there is no specific policy on the use of calculators in middle school, their use is not encouraged in the subject area of mathematics. This is corroborated by the low frequency with which the students responded that they used calculators in mathematics. In addition, in the educational system in Cyprus, there is no unit in the mathematics curriculum that focuses on teaching students how to use calculators. Therefore calculators tend to be used covertly by students who are weak in mathematics.

In the United States, however, there have been decades of research that focused on calculator use. The National Council of Teachers of Mathematics in the United States has even urged the use of calculators to reduce time spent on drill and practice activities and encourage the focus on problems that foster development of underlying mathematical concepts.[34] This research and efforts specifically focused on the proper use of calculators in mathematics instruction have apparently paid off. As a result, the use of calculators is taught in the mathematics curriculum, and their use is permitted in specific units—currently, calculator use is associated with positive results on TIMSS.

This study has demonstrated that although some general patterns in relation to technology use and student achievement exist across countries, differences among the countries can also be found. However, as was the case for calculator use in the United States and Cyprus, policy issues can have a significant impact on how technology is applied in schools. Without policies that are *individualized* to each country's background and needs, the affordances of technology are not always realized. Therefore, the challenge for educators, researchers, and policymakers is to identify the procedures and activities that allow the computer to be used as a cognitive tool that enables students to assimilate knowledge and engage in problem solving via higher-order thinking. These are the skills that could help students increase their achievement in subject areas such as mathematics and science.

Given the constantly increasing percentage of families that own computers, and the decreasing age at which children start to use them, educators need to adapt their practices accordingly. This includes constant evaluation and modification of the way in which computers are used in the classroom as well as the kind of computer-related content that should be taught. However, as was the case with calculators in the United States, research into the subject of educational technology must continue for this tool to reach its full potential. In addition, the professional development of all teachers in educational technology should be ongoing to ensure that technology is used effectively to increase student understanding of course content.

In terms of research, it is suggested that in future cycles of the TIMSS studies, more detailed information regarding computer use should be collected. For example, it would be informative to examine the locations where computer- and calculator-related activities are performed by the students, as well as whether such activities are a required or optional part of their school work. It is also important to determine teachers' rationales for assigning computer and calculator work to their students, and the goals they are trying to achieve through these mediums. Such information could pinpoint the reasons why specific computer-related activities work well for some students in some countries and not as well in other cases. Such information might also help toward bridging the gap between students' home educational background and their school achievement.

Notes

1. Willem J. Pelgrum and Tjeerd Plomp, "Indicators of ICT in Mathematics: Status and Covariation with Achievement Measures," in *Secondary Analyses of the TIMSS Data,* edited by David

F. Robitaille and Albert E. Beaton (Dordrecht: Kluwer Academic Publishers, 2003), pp. 317–30.

2. Seymour Papert, "Situating Constructionism," in *Constructionism*, edited by Idit Harel and Seymour Papert (New Jersey: Ablex, 1991), pp. 1–11.

3. Richard Noss and Celia Hoyles, *Windows on Mathematical Meanings: Learning Cultures and Computers* (Dordrecht: Kluwer, 1996).

4. Susanne Lajoie, Victoria R. Jacobs, and Nancy Lavigne, "Empowering Children in the Use of Statistics," *Journal of Mathematical Behavior* 14 (1995): 401–25.

5. Uri Wilensky, "What Is Normal Anyway? Therapy for Epistemological Anxiety," *Educational Studies in Mathematics* 33, no. 2 (1997): 171–202.

6. Idit Harel and Seymour Papert, "Software Design as a Learning Environment," in *Constructionist Learning*, edited by Idit Harel (MIT Media Laboratory, 1990), pp. 19–50.

7. Penelope H. Dunham and Thomas P. Dick, "Research on Graphing Calculators," *Mathematics Teacher* 87, no. 6 (1994): 440–45.

8. Alan Graham and Michael Thomas, "Tapping into Algebraic Variables through the Graphic Calculator," in *Proceedings of the 21st Conference of the International Group for the Psychology of Mathematics Education*, vol. 3, edited by Erkki Pehkonen (University of Helsinki, 1997), pp. 9–16.

9. Avalos Cedillo, "Algebra as Language in Use: A Study with 11–12 Year Olds Using Graphic Calculators," in *Proceedings of the 21st Conference*, pp. 137–44.

10. Sally Elliott, Brian Hudson, and Declan O'Reilly, "Visualization and the Influence of Technology in A-Level Mathematics: A Classroom Investigation," *Research in Mathematics Education* 2 (2000): 151–68.

11. Chen-Lin C. Kulik and James A. Kulik, "Effectiveness of Computer-Based Instruction: An Updated Analysis," *Computers in Human Behavior* 7 (1991): 75–94; Bruce Rocheleau, "Computer Use by School-Age Children. Trends, Patterns, and Predictors," *Journal of Educational Computing Research* 12, no. 1 (1995): 1–17.

12. See Robert James and Charles Lamb, "Integrating Science, Mathematics, and Technology in Middle School Technology-Rich Environments: A Study of Implementation and Change," *School Science and Mathematics* 100, no. 1 (2000): 27–36; Jay Sivin-Kachala, *Report on the Effectiveness of Technology in Schools, 1990–1997* (Washington, D.C.: Software Publisher's Association, 1998); Gabriela C. Weaver, "An Examination of the National Educational Longitudinal Study (NELS: 88) Database to Probe the Correlation between Computer Use in School and Improvement in Test Scores," *Journal of Science Education and Technology* 9, no. 2 (2000): 121–33; Dale Mann and others, *West Virginia Story: Achievement Gains from a Statewide Comprehensive Instructional Technology Program* (Santa Monica, Calif.: Milken Exchange on Educational Technology, 1998).

13. See Erminia Pedretti, Janice Woodrow, and Jolie Mayer-Smith, "Technology, Text, Talk: Students' Perspectives on Teaching and Learning in a Technology-Enhanced Secondary Science Classroom," *Science Education* 82, no. 5 (1998): 569–89; David E. Shaw and others, "Report to the President on the Use of Technology to Strengthen K-12 Education in the United States: Findings Related to Research and Evaluation," *Journal of Science Education and Technology* 7, no. 2 (1998): 115–26.

14. See Joshua Angrist and Victor Lavy, "New Evidence on Classroom Computers and Pupil Learning," Working Paper 7424 (Cambridge, Mass.: National Bureau of Economic Research, 1999); Elena C. Papanastasiou, Michalinos Zembylas, and Charalambos Vrasidas, "Can Computer Use Hurt Science Achievement? The USA Results from PISA," *Journal of Science Education and Technology* 12, no. 3 (2003): 325–32; Jason Ravitz, John Mergendoller, and Wayne Rush, "What's School Got to Do with It? Cautionary Tales about Correlations between Computer Use and Academic Achievement," paper presented at the annual meeting of the American Educational Research Association, New Orleans, April 2002.

15. Elena C. Papanastasiou, "Factors that Differentiate Mathematics Students in Cyprus, Hong Kong, and the USA," *Educational Research and Evaluation* 8, no. 1 (2002): 129–46.

16. Pelgrum and Plomp, "Indicators of ICT in Mathematics."

17. Harold Wenglinsky, *Does It Compute? The Relationship between Educational Technology and Student Achievement in Mathematics* (Princeton, N.J.: ETS Policy Information Center Research Division, 1998).

18. Ravitz, Mergendoller, and Rush, "What's School Got to Do with It?"

19. John Schacter, "The Impact of Education Technology on Student Achievement. What the Most Current Research Has to Say" (www.mff.org/publications/publications.taf?page=161, October 2006); Rupert Wegerif, "The Role of Educational Software as a Support for Teaching and Learning Conversations," *Computers and Education* 43, no. 1–2 (2004): 179–91.

20. Elena C. Papanastasiou, Michalinos Zembylas, and Charalambos Vrasidas, "An Examination of the PISA Database to Explore the Relationship between Computer Use and Science Achievement," *Educational Research and Evaluation* 11, no. 6 (2005): 529–43; Pelgrum and Plomp, "Indicators of ICT in Mathematics"; Ravitz, Mergendoller, and Rush, "What's School Got to Do with It?"; Wenglinsky, *Does It Compute?*

21. Robert B. Kozma, "ICT and Educational Change. A Global Phenomenon," in *Technology, Innovation, and Educational Change. A Global Perspective,* edited by Robert B. Kozma (Eugene, Ore.: ISTE Publications), pp. 1–10.

22. Ina V. S. Mullis and others, *TIMSS Assessment Frameworks and Specifications 2003,* 2nd ed. (Boston College, 2003), p. 13.

23. Ina V. S., Mullis and others, *TIMSS 2003 International Mathematics Report: Findings from IEA's Trends in International Mathematics and Science Study at the Fourth and Eighth Grades* (Boston College, 2004).

24. Ronald Heck, Terry J. Larsen, and George A. Marcoulides, "Instructional Leadership and School Achievement: Validation of a Causal Model," *Educational Administration Quarterly* 26, no. 2 (1990): 94–125.

25. James L. Arbuckle, Amos 5.0.1 (Chicago: Statistical Package for the Social Sciences, 2003).

26. Barbara M. Byrne, *Structural Equation Modeling with AMOS: Basic Concepts, Applications and Programming* (Mahwah, N.J.: Lawrence Erlbaum Associates Publishers, 2001).

27. Li-tze Hu and Peter M. Bentler, "Cutoff Criteria for Fit Indexes in Covariance Structure Analysis: Conventional Criteria versus New Alternatives," *Structural Equation Modeling* 6, no. 1 (1999): 1–55.

28. Matthew DeBell and Chris Chapman, *Computer and Internet Use by Students in 2003,* NCES 2006-065 (Department of Education, National Center for Education Statistics, 2006).

29. For studies using TIMSS data, see Papanastasiou, "Factors that Differentiate"; Elena C. Papanastasiou, Michalinos Zembylas, and Charalambos Vrasidas, "Reexamining Patterns of Negative Computer-Use and Achievement Relationships. Where and Why Do They Exist?" in *Proceedings of the IRC-2004,* vol. 1: *TIMSS,* edited by Constantinos Papanastasiou (Cyprus University Press, 2004), pp. 127–38. For a study that uses data from the Program for International Student Assessment, see Papanastasiou, Zembylas, and Vrasidas, "Examination of the PISA Database." For a study based on other data sources, see Wenglinsky, *Does It Compute?*

30. James and Lamb, "Integrating Science, Mathematics, and Technology"; Sivin-Kachala, *Report on the Effectiveness of Technology in Schools;* Lani Van Dusen and Blaine R. Worthren, "The Impact of Integrated Learning System Implementation on Student Outcomes: Implications for Research and Evaluation," *International Journal of Educational Research* 21, no. 1 (1994):13–24.

31. Rosamund Sutherland and Nicolas Balacheff, "Didactical Complexity of Computational Environments for the Learning of Mathematics," *International Journal of Computers for Mathematical Learning* 4, no. 1 (1999): 1–26.

32. Wenglinsky, *Does It Compute?*

33. Richard Noss and Celia Hoyles, *Windows on Mathematical Meanings: Learning Cultures and Computers* (Dordrecht: Kluwer, 1996).

34. National Council of Teachers of Mathematics, *Curriculum and Evaluation Standards for School Mathematics* (Reston, Va., 1989).

9

Comparisons between PISA and TIMSS—
Are We the Man with Two Watches?

DOUGAL HUTCHISON AND IAN SCHAGEN

> *A man with one watch always knows what time it is;*
> *a man with two watches is never quite sure.*
>
> —ANONYMOUS

In 2002 and 2003 a strange thing started to happen in the United Kingdom. Government ministers and senior civil servants praised schools' attainment—praise, moreover, based on the results of international studies. The permanent secretary at the Department for Education and Skills wrote, "For those doubters who constantly seek to run down (our education performance), we now have the OECD/PISA study—the biggest ever international study of comparative performance of 15-year-olds in 32 countries—which shows U.K. fourth in science, seventh in literacy, and eighth in mathematics. Only Finland and Canada are consistently ahead of the U.K.—and major countries like Germany, Italy, and Spain are well behind."[1]

Not all of those involved in the political process were equally impressed. Commenting on the results from the 2000 Program for International Student Assessment (PISA 2000), an opposition member of Parliament stated, "It is particularly incredible because in the previous year a far more authoritative study—the third international mathematics and science study, conducted

We are very grateful to the following individuals for encouragement and helpful comments: Lorna Bertrand, Tom Loveless, Scott Murray, Graham Ruddock, Liz Twist, and Chris Whetton. We also would like to thank very warmly all those who responded to our short survey. Not all these wished to be named, but the respondents included Dianne Pennock, Dominique Lafontaine, Inge De Meyer, Ildiko Balazsi, Anna Maria Caputo, Mee-Kyeong Lee, Erna Gille, Marit Kjaernsli, Karl-Goran Karlsson, and Megan Chamberlain.

by the respected International Association for the Evaluation of Educational Achievement—put the U.K. 20th out of 41 countries."[2] The other study he was referring to was the Trends in International Mathematics and Science Study (TIMSS).

So which study is correct? Obviously they are not identical, but how different are they? This chapter describes the differences between TIMSS and PISA in a number of aspects, and also examines the similarities that, at least numerically, seem at first sight to outweigh the differences. In this analysis, the focus on TIMSS is for the eighth grade only.

Frameworks

Following are details concerning the framework for both TIMSS and PISA.

TIMSS

TIMSS is the latest in a line of comparative international studies of attainment performed under the auspices of the International Association for the Evaluation of Educational Achievement (IEA). The IEA describes itself as an independent international cooperative of national research institutions and governmental research agencies.[3] It is essentially a network of education researchers, though with a strong interest in policy. IEA has sponsored surveys in a number of topics, including reading comprehension, French, English, civics, information and communications technology, and literacy, but the focus here is on mathematics and science. The first full study of mathematics took place in 1964, and the second sweep of studies of mathematics and science followed in the 1980s.[4] The Third International Mathematics and Science Study was held in 1994–95, and a repeat was held in 1998–99. Further studies followed on a four-year cycle, with one in 2003 and another being set up for 2007. The acronym "TIMSS" had stuck by this time, and rather than constantly being renamed, the acronym was taken to stand for Trends in International Mathematics and Science Study. Forty-five countries took part in the 1995 TIMSS, thirty-eight in 1999, and forty-nine in 2003.[5] Costs in England are approximately £1 million ($1.9 million).[6]

Over this range of time, IEA has gathered and perfected an enormous amount of knowledge and expertise in the organization, design, administration, and analysis of such studies. IEA and TIMSS are widely respected, rightly, for competence, integrity, innovation, and relevance to the needs of the countries involved. To quote Brown, "The three major international comparisons of mathematics attainment (carried out by) the IEA . . . have had a greater influence on education world-wide than any other single factor during the last 50 year period."[7]

Some of the technical aspects of the study, such as sampling, test design, and analysis, are covered later in this chapter. For now, it is worth paying tribute to the organizational and diplomatic skills involved in keeping on board so many different countries with their own customs, systems, and agendas. In the earlier stages, anecdotal evidence suggested that stringency of standards was sometimes subordinate to ensuring countries' participation. But at least with TIMSS, it is clear that the organizers have been able to insist on rather strict compliance with the rules of the studies.

It is difficult to find a single succinct statement of the aims of TIMSS. This is understandable: IEA studies have been around so long that they collect a range of aims, and they cost so much it is obviously desirable that the subscribers get as much as they can from it. Statements of varying degrees of prolixity can be found in the reports of Robitaille and coworkers and Mullis and colleagues.[8] The former summarizes the kinds of information sought by TIMSS:

—international variations in mathematics and science curricula, including variations in goals, intentions, and sequences of curricula;

—international variations in the training of teachers in science and mathematics;

—the influence of officially prescribed textbooks on the teaching of mathematics and science;

—the course content actually taught in mathematics and science classrooms, that is, opportunity to learn;

—the effectiveness of different instructional practices;

—students' achievement, especially in the area of nonroutine problem solving and the application of science and mathematics in the "real" world;

—the attitudes and opinions of students and teachers;

—the role of technology in the teaching and learning of science and mathematics, particularly the use of calculators and computers;

—participation rates in preuniversity courses, with particular regard to gender-based differences; and

—the effect of tracking, streaming, and other practices used to influence or direct students' course selection.[9]

Interestingly, actual comparison of attainment across countries or over time is not explicitly mentioned above. At least half of the objectives imply that the study aims to account for variation in learning. There is also a strong implication that students should be using mathematics and science as part of their role as a citizen, rather than as part of their employment armory. Concentrating on measuring attainment, Mullis and colleagues state that "it is important . . . that students leaving high school are equipped with a fundamental understanding of science such that the decisions they make are informed ones. . . . Prime reasons for having mathematics as a fundamental part of schooling include the increasing

Figure 9-1. *TIMSS Instruments*

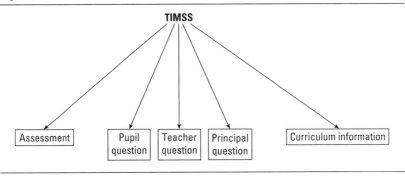

awareness that effectiveness as a citizen and success in a workplace are greatly enhanced by knowing and, more important, being able to use mathematics."[10]

TIMSS is organized around two frameworks: a curriculum framework and an assessment framework. The curriculum framework envisages three layers:

—intended curriculum (what the teacher is expected to teach),

—implemented curriculum (what the teacher taught), and

—attained curriculum (what the pupil learned).[11]

Within the TIMSS assessment instruments (figure 9-1), there are the two subject areas, mathematics and science, plus longer problem-solving questions in mathematics and science from 2003, and questionnaires for students, teachers, and school principals. Each of the two subject areas is further classified into a content dimension (domain) and a cognitive dimension (figure 9-2).

The mathematics content is classified along curriculum lines into number, algebra, geometry, and data and chance, while the science content is classified, again along syllabus lines, into biology, chemistry, physics, and earth science. The cognitive dimension is classified into knowing, applying, and reasoning.[12]

PISA

Compared with the long-established TIMSS series of studies, the Program for International Student Assessment is a relative newcomer. The PISA strategy was defined by participating countries and has a governmental aura since its studies are funded by the Organization for Economic Cooperation and Development (OECD).[13] PISA conducts its surveys on a three-year cycle, with the first in 2000. It is not known precisely at present how much a sweep of PISA costs, but Canada estimated its participation at $1.9 million (approximately U.S. $1.7 million) a year.[14] Costs in England are reported as comparable to those for TIMSS, but slightly higher.[15] Rather than assessing the same subjects as TIMSS, PISA aims to assess literacy in reading, mathematics, and science, which it refers to as domains. The mission statement claims that "the prime

Figure 9-2. *TIMSS Assessment—Content and Cognitive Domains*

Major content domain	Subdomain	Cognitive domains		
		Knowing	Applying	Reasoning
Mathematics	Number			
	Algebra			
	Geometry			
	Data and chance			
Science	Biology			
	Chemistry			
	Physics			
	Earth science			
Problem solving				

Source: Based on David Robitaille and others, *Curriculum Frameworks for Mathematics and Science*, TIMSS Monograph 1 (Vancouver, B.C.: Pacific Educational Press, 1993).

aim of the OECD/PISA assessment is to determine the extent to which young people have acquired the wider knowledge in reading literacy, mathematical literacy and scientific literacy that they will need in adult life."[16]

In each round, one domain is taken as the main subject, occupying about two-thirds of the testing, with the remaining testing time being divided between the other two "minor" domains. Thus in 2000 the main focus was reading literacy; in 2003, mathematical literacy; and in 2006, scientific literacy. OECD also claims that PISA "provides insights into the factors that influence the development of the skills at home and at school and examines how these factors interact and what the implications are for policy development."[17] Therefore, as with TIMSS, there is a suggestion of an attempt to find explanations for differences in performance.

Within the PISA framework, there is the pupil assessment, a student questionnaire, and a school questionnaire for principals (figure 9-3). There is nothing corresponding to the TIMSS curriculum focus. PISA assessments cover three

Figure 9-3. *PISA Instruments*

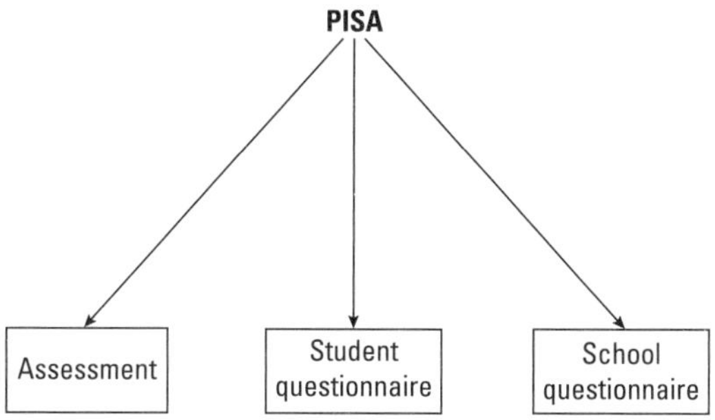

Figure 9-4. *PISA Assessment—Content and Process Domains*

Major content domain	Subdomain	Process domains		
		Knowing	Process	Context (situation)
Mathematical literacy	Quantity			
	Shape and space			
	Change			
	Uncertainty			
Scientific literacy	13 major themes (see text list)			
Reading literacy				
Problem solving				

Source: OECD, *PISA 2003 Assessment Framework*, pp. 24–47.

domains: reading literacy, mathematical literacy, and scientific literacy, plus problem solving (figure 9-4).[18]

Both mathematical and scientific literacy are classified in relation to content (knowing), process, and situations or contexts. "Content" and "process" correspond substantially to TIMSS content and cognitive dimensions.

Mathematical literacy is divided into the "big ideas" of quantity, shape and space, change, and uncertainty, and only secondarily into "curricular strands" such as numbers, algebra, and geometry. The definition of mathematical literacy is "an individual's capacity to identify and understand the role that mathematics plays in the world, to make well-founded judgements and to use and engage with mathematics in ways that meet the needs of that individual's life as a constructive, concerned and reflective citizen."[19] The other big idea that comes across in the documentation is that of "mathematization," meaning converting a question from real life into a mathematical representation and solving and interpreting the resulting mathematical problem.[20]

Scientific literacy is defined as "the capacity to use scientific knowledge, to identify questions and to draw evidence-based conclusions in order to understand and help make decisions about the natural world and the changes made to it through human activity."[21] More extensively, it is "an important life skill . . . the capacity to draw appropriate and guarded conclusions from evidence and information . . . to criticise claims made by others on the basis of the evidence put forward, and to distinguish opinion from evidence-based statements."[22]

Within science literacy, PISA identifies the following major themes:

—structure and properties of matter,

—atmospheric change,

—chemical and physical change,

—energy transformations,

—forces and movement,

—form and function,

—human biology,

—physiological change,

—biodiversity,

—genetic control,

—ecosystems,

—the Earth and its place in the universe, and

—geographical change.[23]

The aims of TIMSS and PISA are explicitly different. Whereas TIMSS focuses on the extent to which students have mastered mathematics and science as they appear in school curricula, PISA aims to capture "the ability to use knowledge and skills to meet real-life challenges."[24] In testimony to the House of Commons Education and Skills Committee, Barry McGaw, then director for education of the OECD, characterized the difference as TIMSS asking, "What science have you been taught and how much have you learned?" while for PISA it was "What can you do with the science you have been taught?"[25]

Sampling

We deal with sampling for each of the two studies in the following.

TIMSS School Sampling

Smithers provides an interesting brief history of the IEA.[26] Each country carried out its own sampling, and in the early days, positive enhancement of the performance of the participating schools and pupils was, allegedly, not unknown. Where a country did particularly well (especially if this was unexpected), on-the-ground knowledge or simply sour grapes led to tales of irregularities in instruction, sampling, and administration of the test material that substantially improved the country's results. At the time, strong political divisions precluded suggestions of in situ checking by unbiased outside personnel, and it is difficult to resist the conclusion that it was considered more important to keep everyone on board and retain them in the sample rather than risk alienating them by close investigation, even had this been possible.

The third round of IEA mathematics and science studies (TIMSS) in 1995 tightened oversight substantially. By 2003 the TIMSS sample was essentially specified by the IEA, with central instructions and specially provided software to perform the sampling. Designs had to be approved by the central coordinator before execution and be carefully documented while being drawn, with details provided to the central site. All this, while doubtless damaging to the amour propre of countries that prided themselves on their sampling skills, knowledge, and experience, was intended to provide a good, reliable sampling result from countries with a range of expertise and manpower at their disposal. The aim was to achieve a good combination of uniformity and sensitivity to local conditions. In this it has been largely successful, except that the more rigid sampling criteria (especially the insistence on a fixed minimum response rate) have disadvantaged those countries (such as England) where there are good school-level data that can compensate for lower response rates. However, this is a debate for another time.

Any country that wanted to participate and could afford the quite substantial costs could be part of the sample. Regions that could be defined as nations within countries (for example, Scotland) or as parts of countries (such as the Canadian provinces) also took part. The target population for the eighth-grade assessment was all students enrolled in the upper of the two adjacent grades that contained the largest proportion of 13-year-olds at the time of testing. This grade level was intended to represent eight years of schooling, counting from the first year of primary or elementary schooling; in most countries, this was the eighth grade.

The international sample design for TIMSS is generally referred to as a two-stage stratified cluster sample design. The first stage consists of a sample of schools, which may be stratified. Schools are sampled systematically from a comprehensive national list of all eligible schools, with probabilities that are proportional to a measure of size, referred to as probability proportional to size (PPS) sampling.[27] The second stage consists of a sample of one or more classrooms from the target grade in sampled schools. For TIMSS 2003, the sample size was defined on the basis of an effective sample size of 400 pupils, after allowing for intracluster correlation. There was also a requirement to conduct analyses at the school and classroom levels, so at least 150 schools had to be selected. Finally, a minimum pupil sample size of 4,000 was set.

Schools could be excluded from the sample if they were in geographically remote regions, were of extremely small size, offered a curriculum or a school structure that was different from the mainstream education system(s), or provided instruction only to students in the categories defined as "within-school exclusions." The last refers to certain types of students who could be excluded, for example, intellectually or functionally disabled students and those who had received less than one year of instruction in the language(s) of the test. Overall, no more than 5 percent of the total pupil population within the cohort could be excluded for the above reasons.

To be placed in the category of "acceptable sampling participation rate without the use of replacement schools," a country had to have both a school response rate *without* replacement of at least 85 percent and a student response rate of at least 85 percent. An overall response rate of 75 percent was required for inclusion in the report. There was provision for replacement schools, and there were some fairly complex arrangements for judging whether the eventual sample was acceptable.

TIMSS Item Sampling

The program aims to assess a wide range of attainments. To cover all these, the pool of items and tasks included in the TIMSS assessment is extensive and would require much more testing time than could be allotted for individual students (about seven hours at grade eight). Therefore TIMSS 2003 and other years used a matrix sampling technique that involves dividing the entire assessment pool into a set of unique item blocks, distributing these blocks across a set of booklets, and rotating the booklets among the students. Each student took one booklet containing both mathematics and science items. This design solicited relatively few responses from each sampled respondent while maintaining a wide range of content representation when responses were aggregated across all respondents. This type of design had been used previously in the England and Wales Assessment of Performance Unit surveys and in the U.S. National Assessment of Educational Progress surveys.[28]

In the TIMSS 2003 assessment design, the 383 eighth-grade items were divided among twenty-eight item blocks. The assessment time for individual students was ninety minutes (six fifteen-minute blocks). The booklets were organized into two three-block sessions (parts one and two), with a break between the parts.

PISA School Sampling

The PISA 2003 sample was largely restricted to OECD member countries, although a number of "partner countries" (for example, Tunisia and Brazil) were also included. The desired PISA target population in each country consisted of 15-year-old students attending educational institutions located within the country, in grades seven and higher. This included full-time and part-time academic and vocational students. Home-schooled and on-the-job students and those not receiving any type of education were excluded. This was unlikely to have posed a major problem in most developed OECD countries, but it could well have been relevant in the "partners." In Mexico and Turkey, the official compulsory school age is from 6 through 14 years old. Even in such highly developed countries such as Germany and Switzerland, schooling is only obligatory up to 15 years of age.[29]

This meant that in all countries testing in April 2003, the national target population was defined as all students born in 1987 who were attending a school or other educational institution. A variation of up to one month in this age definition was permitted. If the testing was to occur at another time, the birth date definition had to be adjusted and approved by the consortium.

The PISA 2003 assessment also used a two-stage stratified sampling design in most countries. A minimum of 150 schools had to be selected in each country. The first-stage sampling units consisted of individual schools having 15-year-old students. In all but a few countries, probability proportional to size (PPS) sampling was used to select schools systematically from a comprehensive national list of all eligible schools. In the selected schools, thirty-five pupils were chosen with equal probability from a list of all 15-year-olds.

Schools could be excluded, for example, on the grounds of inaccessibility or of excluding a language group, possibly due to political, organizational, or operational reasons. Certain types of students also could be excluded, for example, intellectually or functionally disabled students and those who had received less than one year of instruction in the language(s) of the test. A school attended only by students who would be excluded for intellectual, functional, or linguistic reasons was considered a school-level exclusion. It was required that the overall exclusion rate within a country be kept below 5 percent.

A response rate of 85 percent was required for initially selected schools. If the initial school response rate fell between 65 and 85 percent, an acceptable school response rate could still be achieved through the use of replacement schools. In order to compensate for a sampled school that did not participate, where possible two replacement schools were identified for each sampled school. Furthermore, a school with a student participation rate between 25 and 50 percent was not considered as a participating school for the purposes of calculating and documenting response rates. However, data from such schools were included in the database and contributed to the estimates included in the initial PISA international report.

A response rate of 80 percent of selected students in participating (that is, initial or replacement) schools was required. A student who had participated in the original or replacement cognitive sessions was considered a participant. A student response rate of 50 percent within each school was required for a school to be regarded as participating, and the overall student response rate for a country was computed using only students from schools with at least a 50 percent response rate.

Stratification varied among countries; schools were stratified geographically or by such factors as type, size, or academic attainment. Some countries requested some form of control to avoid overlaps between TIMSS and PISA. Sample design was a long and complex procedure, requiring approval from the central organizers. In general, the central body actually drew the sample.

PISA Item Sampling

As in TIMSS, a wide range of items was covered, and matrix sampling was used so that no student was asked to respond to them all. In 2003 the 167 main study items were allocated to thirteen clusters (seven mathematics clusters and two clusters in each of the other three domains), with each cluster representing thirty minutes of test time. The items were presented to students in thirteen test booklets, with each booklet comprising four clusters Each cluster appeared in each of the four possible positions within a booklet exactly once.

Test Items

For both the TIMSS and PISA 2003 studies, the process of item development was long, complex, and thorough, involving a number of different stages, including field trials at which a larger number of items were tested before final item selection.

For PISA 2003, full details of this process can be found in the technical report.[30] The main study contained eighty-five mathematics items, categorized in terms of "competency clusters":

—reproduction, twenty-six items;

—connections, forty items; and

—reflection, nineteen items.

Among the four major content categories, the split was:

—space and shape, twenty items;

—quantity, twenty-three items;

—change and relationships, twenty-two items; and

—uncertainty, twenty items.

Finally, the different item types (question formats) were apportioned as follows:

—multiple choice, twenty-eight items;

—closed constructed response, thirteen items; and

—open constructed response, forty-four items.

For TIMSS 2003, details of item development can be found in Neidorf and Garden.[31] The main study contained 194 mathematics items, of which 115 were new and 79 were "trend" items from previous surveys. The classification of items is similar to but subtly different from that of PISA. In terms of "mathematics cognitive domains," the breakdown of items in 2003 was:

—knowing facts and procedures, twenty-nine items;

—using concepts, thirty-nine items;

—solving routine problems, seventy-eight items; and

—reasoning, forty-eight items.

In terms of the five mathematics content domains in TIMSS, the split was:
  —number, fifty-eight items;
  —algebra, forty-nine items;
  —measurement, twenty-nine items;
  —geometry, twenty-nine items; and
  —data, twenty-nine items.
TIMSS classifies items into only two types. For TIMSS 2003 the breakdown was:
  —multiple choice, 128 items; and
  —constructed response, sixty-six items.

Clearly there are problems in making direct comparisons here between PISA and TIMSS, as the categories defined for the two studies are not the same. However, we have attempted to force such comparisons by conflating certain categories and trying as far as possible to compare like with like. The results of this procrustean procedure are illustrated in figures 9-5 through 9-7 for the three broad classifications into cognitive-competency domains, content domains, and item types.

The "knowing facts and procedures" and "solving routine problems" categories in TIMSS have been combined and compared to PISA's "reproduction" category. Figure 9-5 indicates that this combined category tends to dominate the TIMSS items while the PISA items are more heavily weighted toward "connections" (assumed comparable to "using concepts" in TIMSS). "Reflection" and "reasoning" seem equally weighted in both studies.

Figure 9-6 focuses on the content domains. By using a categorization of PISA items derived by Neidorf and coworkers, we have been able to classify the PISA items into TIMSS categories.[32] TIMSS has a larger proportion of items focusing on the algebra and the number domains, and a smaller proportion focusing on the domain of data and uncertainty.

PISA and TIMSS item types are compared in figure 9-7, with just two categories. The contrast is quite marked here, with PISA favoring constructed response items by two to one, and TIMSS having the same ratio in favor of multiple choice items.

This rather simplistic set of comparisons has led to some interesting conclusions. First, TIMSS emphasizes items that require the reproduction of facts or standard algorithms while PISA focuses on items that demand connections between existing kinds of knowledge. Second, TIMSS has a larger number of items focusing on number and measurement while PISA items are more evenly spread across their content domains. Third, a majority of TIMSS items are multiple choice whereas a majority of PISA items are constructed response.

Figure 9-5. *PISA and TIMSS 2003 Mathematics Cognitive-Competency Domains*

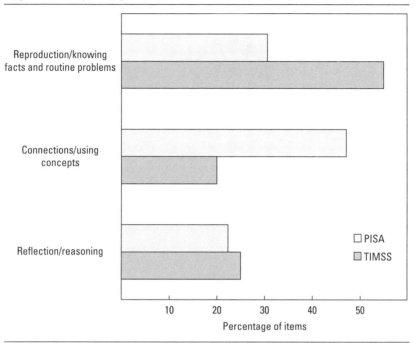

Source: Authors' calculations.

Taken together, these observations seem to imply that PISA items are testing deeper mathematical skills, with a requirement for more integrated thinking and the construction of extended responses. This may be related to PISA's focus on skills for future life rather than on grasp of the school curriculum.

A more in-depth investigation of these issues from an English perspective was undertaken by Ruddock and colleagues, who performed a detailed comparison of mathematics items from PISA, TIMSS, the English national curriculum tests for 14-year-olds, and the General Certificate of Secondary Education examination given to 16-year-olds.[33] With respect to PISA items, they concluded, "It is the quantity of reading that marks PISA out, not the complexity of the language, which is similarly unfamiliar in both the international studies. The high reading demand of questions in PISA is often accompanied by a relatively lower demand in the mathematics or science required. This reflects the lower level of mathematics or science that students can apply in new contexts as opposed to very familiar ones."[34]

Figure 9-6. *PISA and TIMSS 2003 Mathematics Content Domains*[a]

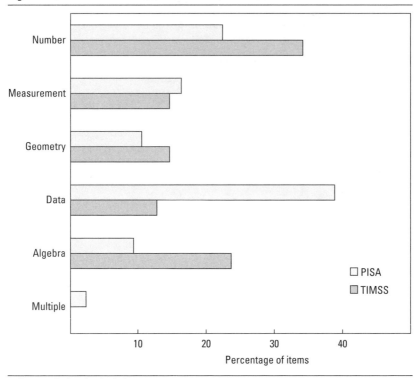

Percentage of items

Source: Authors' calculations.

a. For a definition of "multiple," see Neidorf and others, *Comparing Mathematics Content*, p. 21. No items are classified this way in TIMSS.

Figure 9-7. *PISA and TIMSS 2003 Mathematics Item Types*

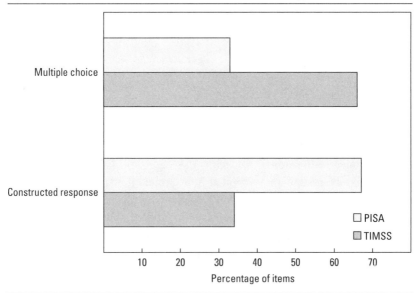

Percentage of items

Source: Authors' calculations.

Use of Item Response Theory

As is generally the case for international studies, pupils' test results for both TIMSS and PISA are analyzed using item response theory (IRT).[35] However, within this general paradigm, there is a certain degree of difference between the two surveys in terms of the type of IRT model used, the number of different scales derived, and how each scale was linked back to the results of previous surveys.

In TIMSS 2003 for the eighth grade, a two- or three-parameter IRT model was used, incorporating a parameter for slope as well as one for item difficulty; a "guessing" parameter was included for the multiple choice questions. The Parscale program was used to fit the data.[36] Separate calibrations were performed, by grouping items appropriately, to generate five mathematics and five science scales. The mathematics scales were number, algebra, measurement, geometry, and data. The results were placed on a metric defined by setting the grade eight results for the countries participating in the 1995 survey to a mean of 500, with a standard deviation of 100.[37]

In PISA 2003, a generalized one-parameter logistic model was used; Con-Quest software was used to fit the model.[38] A total of seven scales were created: one each for reading, science, and problem solving, and four for mathematics (quantity, space and shape, change and relationships, and uncertainty). The results in reading and science were scaled to the metric defined in 2000, but for problem solving and mathematics, new scales were defined, with a mean of 500 and standard deviation of 100.[39]

Despite the aforementioned technical differences between the surveys, there is also a substantial amount of similarity in the techniques used. For example, both surveys generate "plausible values" for each student on each scale to represent the underlying uncertainty or measurement error involved. Typically, five randomly generated values are produced for each, based not only on the IRT model fitted but also conditioned on a range of background characteristics. In principle, secondary analysis of pupil results from these surveys should be based on such plausible values rather than assuming a single outcome for each individual.

In this two questions arise. First, to what extent do the different modeling assumptions in PISA and TIMSS affect the comparability of results from the two studies? Second, how sensitive are these study results in general to modeling assumptions?

Brown and coworkers have attempted to address the first issue, comparing results for four surveys including the 2003 PISA and TIMSS studies.[40] The authors were able to compare results for the 1995 TIMSS study using both a one-parameter and three-parameter model, and concluded from this that cross-country patterns of central tendency are robust to choice of model. However, they concluded that this was not true for measures of dispersion, particularly for less developed countries. This does not directly answer the first question above, but it suggests that some of the different results between surveys may be affected by different modeling assumptions.

To answer the second question about sensitivity to modeling assumptions, we would need to do something rather more radical than has so far been attempted. Although there are some differences between PISA and TIMSS in terms of modeling outcomes, the similarities between them are even more striking. Both are deeply committed to IRT analysis, and there are strong similarities in the derivation of pupil scales and the use of plausible values. For those who are skeptical of the use of IRT, how would we convince them that the substantive findings of these studies are robust and not artifacts of this particular modeling paradigm?

There has been some sensitivity analysis of the impact of changing the details of the modeling on the reported outcomes. The first round of TIMSS was analyzed using the one-parameter logistic model whereas the 1999 round and subsequent sweeps were analyzed using two- and three-parameter models. The results of the 1995 survey were also later reanalyzed using two- and three-parameter models. With one exception, the mean scores on a country basis were found to be essentially the same. However, the dispersion of pupil results does appear to be sensitive to the type of IRT model used.

More generally, the two- and three-parameter models are dependent on rather strong assumptions, and this is also true, a fortiori, for the more restrictive one-parameter approach. There are three important assumptions: unidimensionality, local independence, and universality of application. The results, at least theoretically, are likely to be unreliable if the assumptions do not hold. The analysis teams counter such doubts by extensive testing of the items, but there are still areas of uncertainty.

The first concerns the specifics of testing items. Item fit is tested at the standard 0.05 significance level, which implies that items are rejected as poorly fitting if there is a less than 5 percent probability that the observed difference between model and data could occur by chance. There are some issues here. In a study with 200 items, one would expect about ten to fail this test even if none were truly poorly fitting. In addition, the goodness of fit is a function of both the model and data. In principle and in our experience, one-parameter (Rasch) models tend to be more likely to reject items than the two- or three-parameter alternatives. Thus valid items from a curriculum or assessment perspective can be rejected purely because of model constraints. In this way, model choice can have an impact on item selection and hence indirectly on the whole testing outcome.

The second major concern is over the question of dimensionality. The analysis in both surveys assumes that the items fall on a one-dimensional continuum. From a theoretical, curriculum-oriented view, this is obviously not the case. Algebra is not geometry, biology is not physics, and so on.[41]

Unidimensionality may be acceptable in mathematics, where the subject is generally compulsory and "marching in step" within classes may be sufficient to ensure that everyone learns much the same thing at the same rate.[42] It is obviously going to be less justifiable within science, which consists of a range of much more obviously different topics that are not unidimensional theoretically and may even, in some situations, include alternative subject areas. Some kind of theoretical justification for cramming items into a single dimension is required here, especially since the studies then produce subscales.

The number of items used for an assessment is also something that requires justification. TIMSS 2003 contained 194 items in mathematics: fifty-eight for number, forty-nine for algebra, and twenty-nine each for measurement, geometry, and data. PISA, on the other hand, contained between twenty and twenty-three items in the four main categories for its assessment of mathematics in 2003. For science assessment in 2003, PISA had thirty-five items in total, and the amount of independent information obtained was actually likely to be substantially smaller than this since there were only thirteen distinct question stems.[43]

A similar issue arises with regard to local independence, especially since as noted, PISA's concentration on "realistic" questions means that a relatively small number of topics can be assessed.

Collection of Background Information

Both TIMSS and PISA collect information on the pupils' backgrounds as well as school data, but TIMSS also collects information on teachers. In addition, TIMSS views curriculum aspects as one of the main planks of the study and collects data on this, although this aspect is not discussed here to any great extent.

The rationale given for collecting background information is rather similar in each case. "For a fuller appreciation of what the TIMSS achievement results mean and how they may be used to improve students learning . . . it is important to understand the contexts in which students learn."[44] "Similarly, it was important to ensure that the framework for contextual factors addressed policy-relevant areas and provided the basis for producing internationally significant research."[45] "OECD/PISA . . . provides insights into the factors that influence the development of the skills at home and at school, and examines how these factors interact and what the implications are for policy development."[46]

Table 9-1. *Student and Teacher Questionnaires*

Student or teacher	TIMSS	PISA
Student	Student and home background	Same
	Attitudes to school and school climate	Same
	Attitudes to mathematics	Same
	Mathematics activities in school	Same
	Activities outside school	Learning activities outside school
	Self-rated cognition	Same
	. . .	Learning strategies and preferences
	. . .	Classroom climate
	Computer and technology use	Same
Teacher	Academic preparation and certification	. . .
	Teacher assignment	. . .
	Teacher induction	. . .
	Professional development	. . .
	Teacher characteristics	. . .
	Curriculum topics taught	. . .
	Class size	. . .
	Instructional time	. . .
	Instructional activities	. . .
	Assessment and homework	. . .
	Computers and internet use	. . .
	Calculator use	. . .
	Emphasis on investigation	. . .

Sources: Mullis and others, *TIMSS Assessment Framework and Specifications,* pp. 71–82; OECD, *PISA 2003 Assessment Framework,* pp. 24–47.

Neither assessment actually uses the word "cause" or "causation" in its justification, though PISA perhaps comes rather closer than TIMSS. However, there is a clear implication that there are school and pupil background factors that affect learning, and that information is being collected in order to assess the effects of these factors and how changing them could affect attainment.

Terminology is different, as always, between the two agencies. However, beneath this there is a fair degree of similarity. (See tables 9-1 and 9-2.) Thus TIMSS "home background" corresponds with PISA "student background," and both seek information on such topics as gender, parental education, and employment. The major difference between the two is obviously the teacher questionnaire, administered by TIMSS but not by PISA. TIMSS questionnaires cover both mathematics and science, while PISA deals with only the "lead" topic in each sweep. PISA instruments are also somewhat longer and

Table 9-2. *School Questionnaire*

TIMSS	PISA
School characteristics (location, enrollment, economic background, general atmosphere)	School characteristics (location, enrollment, grade repetition, economic background, instructional hours, limitations to capacity to provide instruction)
Role as principal	Responsibilities in the school
Parental involvement	. . .
Eighth-grade instruction in mathematics and science	Instructional organization for 15-year-olds in mathematics
Information on eighth-grade teachers in your school	Information on teaching staff
Student behavior	Student attitudes
Resources and technology	Funding, computers
. . .	Admissions criteria
. . .	Assessment
. . .	Options for students with other languages

Source: See table 9-1.

therefore typically more detailed. Perhaps the main difference, however, lies in the approach to factors hypothesized to influence learning.

These differences correspond to the frameworks of the two studies. Thus the TIMSS questionnaires concentrate to a large extent on the activities of the schools and teachers in classrooms, and on curriculum. PISA's focus is on individuated characteristics of learners—attitudes toward the subject, learning strategies, and motivation. Both questionnaires include a section on problems that make teaching and learning difficult, such as poor ethos and lack of resources. Interestingly, perhaps none of the questions deals with pluses. "My task is made easier by helpful and knowledgeable colleagues and enthusiastic pupils" is not a statement that appears. It might be useful to consider what this says about the attitudes and assumptions of education professionals worldwide.

The problem with this kind of obsessive data collecting is illustrated by an example from TIMSS. The TIMSS results were chosen because we are more familiar with them, but the point could be equally made with the PISA results.

> One of the main ways for students to consolidate and extend classroom learning is to spend time out of school studying or doing homework. . . . To summarize the amount of time typically devoted to homework in each country . . . TIMSS constructed an index of out-of-

school study time (OST) that assigns students to a high, medium, or low level based on the amount of time they reported studying mathematics, science, and other subjects . . . On average internationally, students at the low index level had lower average mathematics achievement than their classmates who reported more out-of-school study time. However, spending a lot of time studying was not necessarily associated with higher achievement. . . . Students at the medium level of the study index had average achievement that was as high as or higher than that of students at the high level. This pattern suggests that, compared with their higher-achieving counterparts, the lower-performing students may do less homework, either because they simply do not do it or because their teachers do not assign it, or more homework, perhaps in an effort to keep up academically.[47]

Deconstruction of this discussion would go as follows:

—It is well known that doing homework improves performance.

—The more homework pupils do, within reason, the better they should perform. This is not stated directly but is implied by the need to explain away the lack of a regular trend in the results.

—We do not find this at an aggregated level, anyway.

—We shall overrule this apparent finding by arguing the existence of two groups of homework doers: one that performs well because it does homework and another that does homework because it does not perform well.

—There is no evidence in the findings for this. It just based on experience.

While this is a perfectly reasonable interpretation—indeed, one we would probably have made had we been forced to comment on such results—it does very strongly question the relevance of this information. How much has been learned here? In this case, the writers have sensibly refrained from drawing strong conclusions from the results of their analysis. However, there are many more analyses, especially at the national level, where quite unjustified inferences are made from the kind of cross-sectional data available from TIMSS and PISA.

The point here is that there are many examples where an apparent cross-sectional relationship ignores a common cause or the fact that causation can be two way. Thus, in the United Kingdom, pupils in grammar schools perform better on average than those in other types of schools, but this is largely because they are selected to be high performers. Pupils who like mathematics and use certain learning strategies are likely to do well in mathematics, but it is also the case that pupils who do well in mathematics are liable to like mathematics more and to use different learning strategies. Schools that spend more time on task are likely to have higher scores, but conversely, teachers in a class with well-behaved and high-performing pupils are likely to find it easier to spend time on task rather than on low-level control activities. It is always very difficult, if not impossible, to sort out such tangles of influence.

If there are "third factors" that might influence performance, then a longitudinal study collecting information on possible third factors would allow far better answers to research questions. Another point is that the major influence on pupils' performance in these surveys will be prior attainment at an earlier point. The apparent influence of school, class, and pupil background factors can change quite significantly when allowance is made for prior attainment, in what is sometimes known as a value-added analysis. While the logistical difficulties of longitudinal data collection should not be ignored, the gain in terms of worthwhile data would be considerable.

The problems involved in longitudinal data collection in international studies are exacerbated by the need to provide a system that caters to different educational and data collection structures within individual countries. One possibility would be to try to capture (a subset of) the grade-four pupils in TIMSS and match those results to their performance at grade eight. Clearly the grade-eight sampling would need to be modified to include the schools to which these grade-four pupils transferred. However, the major difficulty is probably that of actually tracking pupils through different systems and finding a reasonable number for whom data is available at both time points.

In some educational systems (for example, England), detailed pupil-level information is available nationally, and this tracking process could be done centrally and with (in principle) little effort. In other systems, this is not the case, and more detailed work would need to be done at a local level. For example, for each grade-four school, it should be possible to determine which were the main schools to which pupils transferred by the grade-eight stage. Lists of names could then be supplied to those schools, which could be asked to identify those students who are currently in their eighth grade. In this way, with a nontrivial amount of fieldwork, a good proportion of the fourth-grade pupils could be tracked longitudinally.

An alternative scheme would involve identifying the main grade for the start of secondary schooling for the majority of interested countries and running a special "baseline" assessment for pupils in that grade, to be followed up a year or two later with the main study and linked assessments. Regardless of the approach used, more informative data, based on a significant number of pupils for whom progress over time can be linked to background factors, could permit some extremely useful analyses, and the authors feel strongly that this possibility should be pursued vigorously by the international agencies and governments to add value to the work already being done.

Survey of Countries Involved

Many countries had highly similar rankings in TIMSS and PISA—Japan, Korea, and Italy, for example. Others had distinctly contrasting ratings—for example, New Zealand, Hungary, and Russia when TIMSS 1999 scores were compared with those of PISA 2000, though differences appear to have been less noticeable in 2003.

This led to the question whether there was any obvious reason why some countries had substantially different rankings in the two studies. In particular, we tried to determine whether the reason for the differences was intrinsic—that is, related to the content of the tests—or attributable to other factors, such as the definition of the population or details of sampling or administration. In the latter case, one obvious potential factor lies in how the population is defined, with TIMSS sampling by grade and PISA by age, but there are many other such possibilities.

For this reason, it was decided to contact coordinators in countries that had participated in TIMSS 1999 and PISA 2000, or in TIMSS and PISA 2003. Those who are familiar with international studies may not be surprised that there is some doubt about what constitutes a country or even participation. Thus, for example, Canada participated in PISA, but only some of its provinces took part in TIMSS. England took part in TIMSS 1999, but it was the United Kingdom that participated in PISA 2000. In 2003 England and Scotland took part in TIMSS, but again it was the United Kingdom that participated in PISA—though in the latter instance, the sample from England was judged of insufficient quality to be reported, while the other three constituent countries, including Scotland, were reported individually.[48]

From the websites of TIMSS and PISA, we were able to identify the countries participating and then the overlap.[49] We e-mailed coordinators from twenty-three countries and received replies of varying degrees of extensiveness from twelve of them; thus any statistical description of replies should be taken as indicative rather than representative. There were two main strands to our inquiry. First, we sought any publications comparing the two programs. Second, we asked a number of questions, mainly about the administration and design of the two studies.

While individual respondents gave careful and extensive answers to the questions, and many countries produced descriptions of the import of TIMSS or PISA results to their schools, there was a surprising lack of officially commissioned comparisons of the results of the two studies. England produced a comparison between TIMSS and PISA items and those in the public examination and assessment system, and the United States produced documents in which the TIMSS items and PISA items were separately compared to those of the National Assessment of Educational Progress.[50] Ireland also produced a comparison report mainly looking at PISA 2000 with references to TIMSS 1995, but otherwise little else on this topic seems to have been published.[51] The following lists the survey questions and summarizes their principal responses.

Were There Any Differences in Your Country between the Results of TIMSS and PISA?

Of the twelve respondents, three considered that their PISA results were better, one indicated that TIMSS results were better, four reckoned that there were no or slight differences, one stated that there were differences without specifying what they were, and one was unable to answer the question as TIMSS did not survey the entire country. One country's respondent stated that TIMSS had shown a decline in performance but PISA had not, and another offered the comment that while there was no overall difference, the differences in male-versus-female performance were larger in TIMSS.

If There Were Any Such Differences, Were You Surprised by Them?

Only one country's respondent indicated surprise at the differences.

In Your Opinion, Were Differences in the Samples Likely to Give Rise to Differences in the Results?

None of the respondents expressed the opinion that differences in the samples were likely to give rise to differences in the results, except that they obviously related to different age groups.

In Your Opinion, Were Differences in the Instructions to Schools Likely to Give Rise to Differences in the Results?

Of the ten country representatives who replied, none considered that differences in the instructions to schools were likely to give rise to differences in the results. However, not all coordinators were familiar with the details of instructions for both studies.

In Your Opinion, Were Differences in the Response Rates Likely to Give Rise to Differences in the Results?

None of the respondents considered that differences in the response rates were likely to give rise to differences in the results.

To summarize, the main finding from this brief survey is that most respondents considered it unlikely that crude differences in the administration details generated any observed discrepancies between TIMSS and PISA. Many of the differences appeared to arise as a result of the age definition of the population, either directly or because of curriculum-age interactions or differences in repeating grades. In some countries, it was suggested that performance comparisons were in some way biased because operationalization of the grade definition led to differences in the age of the tested population. Thus, for example, as a result of the grade definition details, the average age of pupils tested for TIMSS 1995 in Scotland was lower—in some cases up to a year lower—than that of pupils in other countries, for example, Germany. This was not the case in PISA since it had an age-based sample, though conversely, it would mean that some students were likely to have had more time at school.

In some countries, it appeared that there was an interaction between age and curriculum effects. Thus, in Belgium (Flanders), it was suggested that at grade eight (TIMSS), Flemish pupils are not taught chemistry and geology, while by grade ten (PISA), most have been introduced to chemistry and the topics tested in geology.

The effects of specific examination or testing programs at various ages were also put forward by some of the coordinators as a reason for differences in TIMSS-versus-PISA rankings. Hungary, for instance, has an entrance examination for upper secondary school at the end of grade eight, and it is possible that heightened exam awareness is likely to increase student achievement in TIMSS. Conversely, the absence of exams may also be a factor. According to New Zealand school coordinators, TIMSS 2003 was the first formal assessment of long duration that some students had experienced. Similarly, Scottish pupils may be less used to formal assessment than students in other countries since there are no national tests or new national assessments in science in Scotland.

As intimated above, national policies on repeating years differ between countries. At the one extreme, England's pupils mainly move up through the schools by age, to such an extent, in fact, that there are not any statistics about this. In other systems, there is a sizable amount of grade repetition. In Flemish-speaking Belgium, for example, 72.2 percent of those tested in PISA were in grade ten, 22.8 percent were in grade nine, and 2.5 percent were in grade eight.

The discussion so far has looked at obvious structural factors that could confuse inferences based on the overt performance data from TIMSS and PISA. We now turn to a direct comparison between those results. It is important to bear in mind that differences in scores between TIMSS and PISA do not in themselves represent differences in absolute performance. Even were the two tests comparable, the students in the PISA sample are substantially older than those in the TIMSS sample, although each test is standardized to an international mean of 500. The figures shown below should give an impression of the relative ordering of countries. Results have been reported in original country scores, rather than rankings, to establish the relative sizes of differences.

Figure 9-8. *Mathematics Scores, PISA versus TIMSS, 2003*[a]

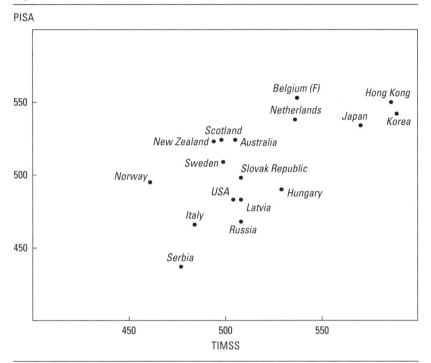

Source: OECD, *Learning for Tomorrow's World: First Results from PISA 2003*; Martin, Mullis, and Chrostowski, *TIMSS 2003 Technical Report*, exhibit 1.1.
a. (F) refers to Flanders. Two "partner" countries omitted for clarity.

Figures 9-8 and 9-9 show the mean PISA scores graphed against the TIMSS scores for countries participating in both surveys in 2003. Some differences are not particularly easy to see, so figures 9-10 and 9-11 have been provided to show the residuals generated when predicting PISA scores from TIMSS, which may give a clearer impression of relative PISA-versus-TIMSS differences.

Bearing in mind all the caveats and possible alternative explanations just described, there is a clear suggestion of a pattern, with the "first world" countries generally doing better and former Warsaw Pact and Far Eastern countries generally doing worse on PISA than would be predicted on the basis of their TIMSS performance. Interestingly, the Hungarian national coordinator had this to say about the Hungarian situation: "TIMSS focuses on the curriculum-related tasks, while PISA is literacy based. [The] Hungarian school system still highly relies on factual knowledge and traditional teaching strategies, so

Figure 9-9. *Science Scores, PISA versus TIMSS, 2003*[a]

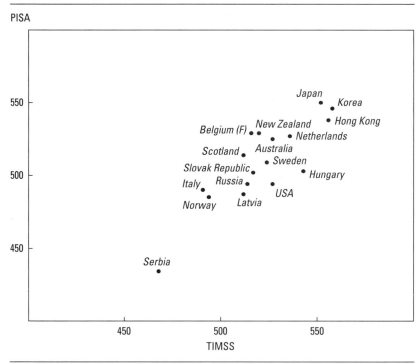

Source: See figure 9-8.
a. (F) refers to Flanders. Two "partner" countries omitted for clarity.

students are relatively good in tasks which are close to their usual classroom tasks, while they meet relatively few literacy-based tasks, and they do not know what to do with these."[52]

Similar results emerged when we compared TIMSS 1999 and PISA 2000. It seems that the PISA assessment may be more aligned with the education systems of western European countries than the TIMSS assessment is.

Conclusions

Do these surveys in fact represent two distinct exercises? The rhetoric would suggest so. There is "mathematics literacy" versus "mathematics" and "science literacy" instead of "science." And there is an apparent difference in the philosophy of each approach. If one can sum up the differences in a metaphor, TIMSS is inside the school wondering what makes it tick while PISA is outside on the street waiting to see what is coming out. PISA also seems to have a more individual-oriented approach to the data: where TIMSS has detailed

Figure 9-10. *PISA Mathematics 2003 Residuals (from TIMSS 2003) versus TIMSS Mathematics 2003*[a]

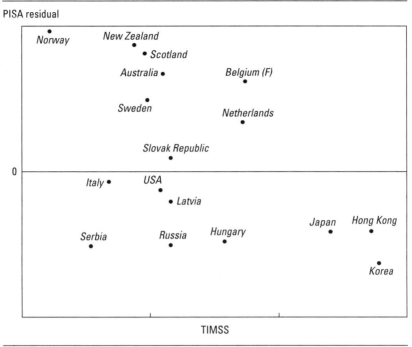

PISA residual

TIMSS

Source: Authors' calculations.
a. (F) refers to Flanders. Two "partner" countries omitted for clarity.

investigations of what is taught and how, PISA's angle is to compare the learning styles and motivations of the individual pupils.

So is this expressed in an "on-the-ground" difference between the two studies? If they were, in reality, very similar, then simply giving them different labels would not make them different. As discussed earlier in this chapter, there are actually a great number of similarities between them. In fact, the PISA sampling and analysis appear to have been largely lifted from the TIMSS model, refined as it has been over years of experience. To take just a few examples at random, both studies involved about 4,000 pupils, used a PPS sample of schools, employed a matrix sampling design, and analyzed the data using IRT and plausible values. And there are more similarities. To be sure, PISA tested students at a different stage of education, was based on a year group rather than a grade group, sampled classes rather than individual pupils, and used the one-parameter model rather than a two- or three-parameter model. Yet these are not essential differences, although they may generate some alterations in the rank ordering of countries, as discussed earlier.

Figure 9-11. *PISA Science 2003 Residuals (from TIMSS Science 2003) versus TIMSS Science 2003*[a]

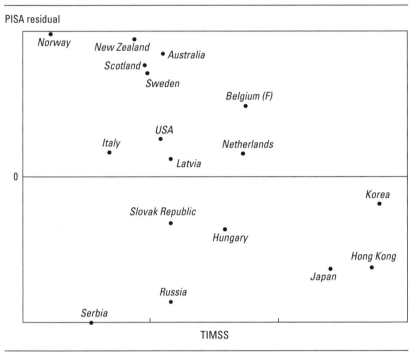

PISA residual

Source: Authors' calculations.
a. (F) refers to Flanders. Two "partner" countries omitted for clarity.

The main difference lies in the type of item. PISA items are aimed at life skills while TIMSS items are more knowledge oriented. However, to quote Smithers, "Just because it says it's curriculum-free, doesn't necessarily mean it is."[53]

Where TIMSS questions are more direct and abstract, PISA questions are more lengthy and wordy. TIMSS asks, "What do you know?" while PISA asks, "What can you do?"

Some differences in results appear to be based substantially on the differences in the interaction between age and curriculum. For example, students from Flemish Belgium appear to perform relatively less well at age 13 (TIMSS) because they have not generally been introduced to algebra, but they catch up by age 15. However, we appear to have established through our small survey of coordinators that, otherwise, sample definition, response rates, or other technical questions do not seem to affect the national scores differentially between the two studies. Even after allowing for possible alternative explanations, there seems to be a pattern that highly developed, Western countries do relatively well on PISA while the former Warsaw Pact countries perform less well. This, in turn, seems to correspond to how these countries conduct their respective education systems, with the former Warsaw Pact countries concentrating on a more formal, traditional approach while the Western countries focus on interpretation and application.

Thus there do seem to be some differences and probably ones with worthwhile aspects. It is useful to know if the education system is working. It is also useful to know what the graduates of the system can actually do. However, if the two studies are meant to be different, shouldn't they really differ a bit more than they do? Life skills are presumably the ability to hold down a job, engage in basic financial dealings, and make political judgments based on intelligent and informed assessment of the claims of politicians. If this is what PISA is supposed to be measuring, why is it conducting its assessments in schools, at a stage before most pupils leave education, and using a paper-and-pencil test administered to pupils sitting in rows in classrooms? And what evidence is there that the items provided really reflect life skills?

So, there are two comparative education survey systems—but does the world actually need them both? Perhaps before asking this, the first question should be whether the world actually needs even one? The first and most noticeable outcome of these studies is the "league table" of countries, which always nowadays creates a stir, especially if the country concerned has been unexpectedly unsuccessful. In our opinion and that of educational experts, too much is made of the league table rankings. To be fair, the survey systems are aware of this and appear to concur with this view. PISA, for example, gives the ordering of its results in statistically comparable groups. However, in our opinion, these are based on estimated standard errors that are too small as they take no account of item sampling, a potentially serious problem in PISA with its relatively small number of items. There certainly is a potential problem of overinterpretation—particularly biased interpretation—of the results. Smithers sums this up in a delightful sentence: "The cloud of data generated becomes a canvas on to which the committed can project what they want to see."[54]

In modern cultures where governments feel impelled to manage societal functions, it seems useful to ensure that decisions are based on the "right" information, and international studies do offer the possibility of comparative information that would not be obtainable any other way. For example, for the United Kingdom, these surveys provide the only internationally comparable source for the output of the four different education systems within the country.[55] More generally, the surveys give countries trying to plan their future an indication of the strengths and weaknesses in their skills base.

However, this is a downside. These studies are expensive to conduct: TIMSS and PISA each cost participating countries approximately $2 million.

Even more to the point is the degree of demand placed on already over-burdened schools, especially at the secondary school level.[56]

Examination and testing are social products and as such evolve as society changes. Up to and immediately after World War II, Western society was hier-archical, rigid, and unchanging, with relatively few desirable occupations. There was relatively little change in the information available, and the main aim of testing was to order the candidates according to their skills and use this to guide and justify their entry into various vocations. More recently, with the accelerating expansion of knowledge and the widening range of opportunities, the focus in education has shifted to ensuring the maximum effective acquisition of particular skills. With the advent of globalization, outsourcing, and computerization, the nature of employment has changed greatly; in developed countries, it has become more important to be able to analyze a situation and specify what is to be done than it is to carry out the action. Concurrently, the democratic process, or at least the ability of citizens to have an input in the running of their society, has expanded in a very large number of countries. And this, in turn, points to the importance of having a citizenry that can assess to some extent the validity of the statistical basis for political arguments. This implies that the PISA goal of measuring literacy in the areas concerned is an important one.

Both TIMSS and PISA set themselves up to measure change over time, the former since 1995 and the latter since 2000. Indeed, the name *Trends in Mathematics and Science Study* could be taken as a bit of a clue to this orientation. While this is a strength, it is also a source of vulnerability, and a study can become a prisoner of its past. It is well known that if one wants to measure change, then one should not change the measure. Yet with TIMSS the measure cannot remain the same because at least some of the items are released after each sweep. The use of the Rasch or IRT models is intended to deal with this. However, even if this is considered possible, it can only occur if the tests are measuring the same thing. Conceptually and in practice, the PISA items have a rather different orientation than those in TIMSS. If TIMSS wants to add the assessment of life skills to its mission, it will find it difficult to fulfill this goal using its trends in measures up to this point. And PISA will undoubtedly face the same challenge in the future when the focus in education and society moves on. How do such studies maintain links with what has come before while also ensuring that test practice and orientation remain as current as possible? Do the studies keep adding new tasks, eventually becoming so unwieldy that they sink under their own weight? Or do they simply jettison the old parts and lose continuity with the past? In our opinion, the need for comparability over long periods of time is less important

COMPARISONS BETWEEN PISA AND TIMSS   257

than one might imagine. At present there are so many changes in curriculum and emphasis that time series more than 10 years old are really not very informative.

Questions

Our evaluation and comparison of TIMSS and PISA has engendered many new questions, ones that could merit further research. The two studies differ in the intensity of the coverage and the number of items, with TIMSS having over twice the number of items than in a "main" PISA year and five times the number in a "minor" PISA year. The effective number of items is actually even smaller because of the PISA habit of embedding a number of questions within a single question stem. While this improves realism, it also potentially reduces generalizability. Can PISA justify an item sample this small, and are the minor PISA years of any value?

What is the theoretical (as opposed to the practical) justification for using any kind of IRT model, with its unidimensionality assumption, on such a heterogeneous topic area as science?

How sensitive are the main outcomes of the studies to the different models used for analysis?

Would it be possible to introduce some kind of computer-adaptive testing to decrease the volume of assessment needed to achieve a given degree of precision?

There seems to be little formal research into the reasons for the apparent differences in the relative positions of countries in the two studies. Why is such research not a matter of the highest priority for both agencies?

Recommendations

The apparent competition between TIMSS and PISA highlights the importance of continuing development both of goals and techniques. It is arguable that PISA has stolen a march on TIMSS by introducing the life skills aspect to evaluation. This innovation demonstrates the importance of ongoing methodological development as part of any such study.

It seems unfortunate that there are two large international studies in the field at the same time doing very similar things. As such, they are bound to view themselves in competition. Although competition can sharpen up an outfit's operation, it can also have unfortunate consequences. PISA was able to hit the ground running largely because it lifted its methodology from TIMSS. The IEA, in the best academic tradition, has always been extremely open in allowing inspection of its methodology, but after this experience, one would hardly blame the association if it were to shield its work under the guise of commercial confidentiality. Such a reaction would not only impede the development of research capacity in many countries but also make it less likely that methodology would be tested and proved in the normal academic process.

In light of these issues, we present a number of recommendations. First, both the IEA and OECD should maintain open access to their methodologies, and encourage criticism and debate from the wider academic community. This would help to dispel the impression that they operate like a "closed shop" and should be done in a spirit of openness and willingness to learn and improve, recognizing that there is not necessarily a "correct" answer to each technical problem.

Second, more analysis should be done on the sensitivity of the main survey results to the exact details of the models used (for example, one-parameter versus two-parameter IRT). Such sensitivity analyses could be carried out by third parties using the raw data.

Third, the importance of comparing attainment over long time periods should be downgraded, to a maximum of 10 or possibly 12 years, since education curricula and aims change so rapidly that old results are no longer relevant. This may affect the frequency of studies and also the extent to which common test items are retained through several surveys.

Fourth, there should be careful, in-depth investigations of the apparent discrepancies between TIMSS and PISA results at the country level.

Results from cross-sectional studies do not generally allow for a valid impression of the effect on achievement of any aspect of school structure or practice, or pupil attitudes or learning styles. Therefore, the feasibility of introducing a longitudinal study, following the same pupils over a period of years, should be actively investigated. While recognizing the logistical challenges, we believe the potential benefits, in terms of increased understanding of educational systems, would be significant.[57]

Finally, the IEA has achieved an enviable reputation for scholarship, research, and integrity over its long and distinguished career. It is important for the future of international educational research that the association preserve this high standard, and we believe that this will be served best by maintaining its independent, nongovernmental status.

Notes

1. Sir David Normington, "Transforming Secondary Education," *Education Journal,* no. 62 (2002): 6–8.
2. Nick Gibb, "Education, Culture, Media and Sport," 15 November 2002, *Parliamentary Debates,* Commons, vol. 394 (2002), cols. 320–25 (www.publications.parliament.uk/pa/cm200203/cmhansrd/vo021115/debindx/21115-x.htm).

3. International Association for the Evaluation of Educational Achievement, "Mission Statement" (www.iea.nl/mission_statement.html).
4. Alan Smithers, *England's Education. What Can Be Learned by Comparing Countries?* (University of Liverpool, 2004).
5. Ibid. See also Ina Mullis and others, *TIMSS 2003 International Mathematics Report* (Boston College, 2004), p. 15.
6. Personal communication from Lorna Bertrand, chair, PISA National Committee, U.K., 2006.
7. Margaret Brown, "Problems of Interpreting International Comparative Data," in *Comparing Standards Internationally: Research and Practice in Mathematics and Beyond,* edited By Barbara Jaworski and David Phillips (Oxford: Symposium Books, 1999), pp. 183–205.
8. See David Robitaille and others, *Curriculum Frameworks for Mathematics and Science,* TIMSS Monograph 1 (Vancouver, B.C.: Pacific Educational Press, 1993); Ina Mullis and others, *TIMSS 2007 Assessment Frameworks* (Boston College, 2005).
9. Robitalle and others, *Curriculum Frameworks.* There is a comparable statement in Mullis and others, *TIMSS 2007 Assessment Frameworks.*
10. Ina Mullis and others, *TIMSS Assessment Frameworks and Specifications 2003* (timss.bc.edu/timss2003i/frameworksD.html).
11. This has no counterpart in PISA.
12. Michael Martin, Ina Mullis, and Stephen Chrostowski, eds., *TIMSS 2003 Technical Report* (Boston College, 2004).
13. Organization for Economic Cooperation and Development (OECD), *Measuring Student Knowledge and Skills—A New Framework for Assessment* (Paris, 1999).
14. Council of Ministers of Education, Canada, OECD, Programme for International Student Assessment (PISA), and Youth in Transition Survey, "Fact Sheet," December 2001 (www.cmec.ca/pisa/2000/factsheet.en.pdf).
15. Personal communication, Lorna Bertrand, 2006.
16. OECD, *Learning for Tomorrow's World: First Results from PISA 2003* (Paris, 2004), p. 4.
17. OECD, *The PISA 2003 Assessment Framework—Mathematics, Reading, Science and Problem Solving Knowledge and Skills* (Paris, 2003), p. 10.
18. The problem-solving component was introduced in 2003.
19. OECD, *PISA 2003 Assessment Framework,* p. 24.
20. Ibid., p. 26.
21. Ibid., p. 15.
22. OECD, *Measuring Student Knowledge and Skills,* p. 59.
23. OECD, *PISA 2003 Assessment Framework,* p. 136.
24. OECD, *Learning for Tomorrow's World.*
25. Barry McGaw, "Minutes of Evidence. Examination of Witnesses (Questions 1–15)," House of Commons, Education and Skills Committee, March 20, 2002 (www.publications.parliament.uk/pa/cm200102/cmselect/cmeduski/711/2032002.htm).
26. Smithers, *England's Education.*
27. PPS sampling produces a sample that is unbiased at the pupil level with respect to school size when a fixed number of pupils are chosen per school.
28. See Derek Foxman, Dougal Hutchison, and Barbara Bloomfield, *The APU Experience: 1977–1990* (London: School Examinations and Assessment Council, 1991); Albert Beaton, ed., *Expanding the New Design: The NAEP 1985–86 Technical Report,* 17-TR-20 (Princeton, N.J.: National Assessment of Educational Progress, 1988).
29. See Sig Prais, "Cautions on OECD's Recent Educational Survey," *Oxford Review of Education* 29, no. 2 (2003): 139–63.
30. OECD, *Learning for Tomorrow's World,* pp. 15–32.
31. Teresa Neidorf and Robert Garden, "Developing the TIMSS 2003 Mathematics and Science Assessment and Scoring Guides," in *TIMSS 2003 Technical Report,* edited by Martin, Mullis, and Chrostowski, pp. 23–66.

32. Teresa Neidorf and others, *Comparing Mathematics Content in the National Assessment of Educational Progress (NAEP), Trends in International Mathematics and Science Study (TIMSS), and Program for International Student Assessment (PISA) 2003 Assessments,* Technical Report NCES 2006-029 (Department of Education, National Center for Education Statistics, 2006), p. 78.

33. Graham Ruddock and others, *Validation Study of the PISA 2000, PISA 2003 and TIMSS-2003 International Studies of Pupil Attainment,* Research Report 772 (London: Department for Education and Skills, 2006).

34. Ibid., p. 123.

35. For a description of such models, see David Thissen and Howard Wainer, *Test Scoring* (Mahwah, N.J.: Lawrence Erlbaum, 2001), or Robert Hambleton, Harihan Swaminathan, and Jane Rogers, *Fundamentals of Item Response Theory* (Newbury Park, Calif.: Sage, 1991).

36. Parscale, Ver. 4 (Groningen, Netherlands: Science Plus Group, 2003).

37. Eugenio Gonzalez, Joseph Galia, and Issac Li, "Scaling Methods and Procedures for the TIMSS 2003 Mathematics and Science Scales," in *TIMSS 2003 Technical Report,* edited by Martin, Mullis, and Chrostowski, pp. 253–74.

38. ConQuest (Horwich, U.K.: 1997). The one-parameter logistic model was referred to in the report as a Rasch model.

39. See OECD, *Learning for Tomorrow's World.*

40. Giorgina Brown and others, "Cross-National Surveys of Learning Achievement: How Robust Are the Findings?" S3RI Applications and Policy Working Paper (University of Southampton, 2005).

41. For a detailed discussion of this issue, see Bill Schmidt, Robert Houang, and Curtis McKnight, "Value-Added Research: Right Idea but Wrong Solution?" in *Value Added Models in Education: Theory and Applications,* edited by Robert Lissitz (Maple Grove, Minn.: JAM Press, 2005), pp. 145–64.

42. Even within mathematics, there was some evidence in the U.K. Assessment of Performance Unit surveys for differential rates of change between subjects over time. See Foxman, Hutchison, and Bloomfield, *The APU Experience.*

43. OECD, *Learning for Tomorrow's World,* p. 419.

44. Mullis and others, *TIMSS Assessment Framework and Specifications 2003,* p. 81.

45. Mullis and others, *TIMSS 2007 Assessment Frameworks.*

46. OECD, *PISA 2003 Assessment Framework,* p. 10.

47. Ina Mullis and others, *TIMSS 1999 International Mathematics Report,* 2000 (timss.bc.edu/timss1999i/math_achievement_report.html), p. 123–24.

48. Scotland and Northern Ireland had national reports. Wales did not have a separate sample, so it was represented by only a very small number of schools. In the international report, the overall results for Scotland, Northern Ireland, and (erroneously) Wales were reported in one of the appendixes.

49. See "Countries Participating in TIMSS 2003" (timss.bc.edu/timss2003i/countries.html), and "Programme for International Assessment (PISA): Participating Countries 2003" (www.pisa.oecd.org/pages/0,2966,en_32252351_32236331_1_1_1_1_1,00.html).

50. See Ruddock and others, *Validation Study;* Neidorf and others, *Comparing Mathematics Content;* National Center for Education Statistics, *Comparing NAEP, TIMSS, and PISA in Mathematics and Science* (nces.ed.gov/TIMSS/pdf/naep_timss_pisa_comp.pdf [November 2006]).

51. Graham Shiel and others, *Ready for Life? The Literacy Achievements of Irish 15-Year Olds with Comparative International Data* (Dublin: Educational Research Centre, 2001).

52. Personal communication from Ildiko Balazsi, coordinator, TIMSS and PISA, Hungary, 2006.

53. Smithers, *England's Education.*

54. Ibid.

55. Or at least they would if all four systems took part successfully in the same study at the same time.

56. It is a point of interest that the very first piece of education research done by one of the authors in the early 1970s was to investigate an accusation that schools were overloaded with requests for statistical information, a complaint sparked by none other than one of the first IEA studies. *Plus ça change.*

57. An example of combining data from an international study with a longitudinal investigation—albeit for a different age than what we have proposed and with education scores as a baseline rather than an outcome—comes from the Longitudinal Surveys of Australian Youth. These are a series of surveys, from 1995 onward, of the respondents in the PISA study, extending in age from 15 to about 25 years old. The surveys are intended to provide a description of pathways into work and of factors that influence how successful this transition is. See Robyn Penman, *What Do We Know about the Experiences of Australian Youth?* (Camberwell, Australia: Australian Council for Educational Research, 2004).

Contributors

JAN-ERIC GUSTAFSSON
University of Gothenburg

LAURA S. HAMILTON
RAND Corporation

RICHARD T. HOUANG
Michigan State University

DOUGAL HUTCHISON
*National Foundation for
  Educational Research*

JEREMY KILPATRICK
University of Georgia

MICHAEL O. MARTIN
Boston College

JOSÉ FELIPE MARTÍNEZ
University of California–Los Angeles

VILMA MESA
University of Michigan

INA V. S. MULLIS
Boston College

ELENA C. PAPANASTASIOU
Intercollege, Cyprus

EFI PAPARISTODEMOU
Ministry of Education, Cyprus

IAN SCHAGEN
*National Foundation for
  Educational Research*

WILLIAM H. SCHMIDT
Michigan State University

GABRIELA SCHÜTZ
Ifo Institute for Economic Research

FINBARR SLOANE
Arizona State University

Index

Achievement and achievement levels: age of students and, 3, 42, 45–52, 54–55, 61, 148, 250; attendance and, 42; coherent development and, 66; cultural support for mathematics, 2; curriculum and, 156; class size and, 3, 42–43, 52–59; educational focus and, 2; educational tracking and, 44, 87, 91, 115–16, 122, 229; gender differences, 148; home environment and, 2, 32–33; mathematics reforms and, 5; measures of, 156–58; school size and, 6, 52–59, 175–203; school tracking and placement and, 44, 87, 91, 115–16, 122, 229; socioeconomic differences, 6, 53, 54, 148, 200–202, 213; teacher variables, 30–31; technology use and, 6, 205–23; wealth and, 2. *See also* Mathematics achievement

Achilles, Charles M., 58

Age and age differences. *See* Achievement; Grade level; Mathematics achievement; Students and learners

Algebra: achievement in, 149, 151–52; cognitive demand of, 98, 99, 100, 107–08, 109–10; content and process of, 118–20; content and representation, 96–97, 98–99, 106–07, 108–09; curriculum of, 143; functional thinking and, 122–23; graphic calculators and, 207; high-performance items, 95–101; history of algebra in the U.S., 87–88; item characteristics, coding, and comparison, 92–95; low-performance items, 106–18; relative high-performance items, 101–06; relative low-performance items, 111–18; teaching of, 159; textbooks and curriculum, 4, 86; transition to algebraic thinking, 104; U.S. students' performance in, 89–123. *See also* Geometry; Measurement; Patterns, relations, and functions

Allardt, Erik, 38

Angrist, Joshua D., 53–54

APU. *See* Assessment of Performance Unit

Patterns, relations, and functions (mathematical): NCTM standards and, 82, 83; TIMSS and, 69, 89–90, 92–93, 96, 100, 101, 104, 112, 118, 120, 121; U.S. textbooks and, 4. *See also* Algebra

Pelgrum, Willem J., 208

Philippines, 30, 32, 33, 191, 201

Physics, 29

PIRLS. *See* Progress in International Reading Literacy Study

PISA. *See* Program for International Student Assessment

Plomp, Tjeerd, 208

Population issues, 54

Probability, 91

Program for International Student Assessment (PISA): age and, 250; analysis of, 241–43; collection of background information, 243–47; cost of, 230; history, framework, and goals of, 7, 38, 86, 230–33, 227, 245; items of, 237–40, 243, 244t, 254, 257; sampling for, 235–37, 253; TIMSS and PISA comparisons, 227–58; U.K. scores on, 227–28; U.S. scores on, 121–22

Progress in International Reading Literacy Study (PIRLS), 61

Quebec (Canada), 101, 200, 201

Racial and minority issues, 58

Raizen, S. A., 143

Ravitz, Jason, 208–09

Reading: class size and, 53, 58; PISA and, 7, 230–32, 239, 241; teaching and, 39; TIMSS and, 37–38

Reforms: *1990s*, 5, 127–28; educational progressivism, 5; instructional practices and topics covered, 136–38, 140–48, 156; mathematics achievement and, 5, 148–52, 157, 158; multilevel modeling of student

reports, 138–39; reform versus traditional practices, 128–48; research on reforms, 130–31, 157–58; student-teacher agreement, 144–48; TIMSS data and analysis of, 5, 131–40, 158. *See also* Curriculum; Teachers and teaching; *individual countries*

Robinson, Glen E., 53

Robitaille, David, 229

Romania, 46

Ruddock, Graham, 239

Rural areas, 54–55

Rush, Wayne, 208–09

Russian Federation: computers and calculators and achievement in, 6, 210, 214–21, 222; curriculum issues in, 29, 30; language effects in, 199; mathematics achievement in, 25; PISA scores in, 248; TIMSS scores in, 25, 248

Saudi Arabia, 2, 200, 201

Schagen, Ian, 7, 227–61

Schmidt, William, 3, 4, 65–84, 86, 120, 143

Schools: choice of, 181; costs of, 175; implemented curriculum and, 11; effects of autonomy of, 177; effects of class size, 3, 42–43, 52–59; effects of homework, 44; effects of school size and enrollment, 5–6, 175–203; safety of, 34; school-leaving age cohort, 29; school organization and structure, 27–29; student performance and, 42; survey on PISA and TIMSS, 248–52; testing programs of, 250. *See also* Grade level; Students and learners; Teachers and teaching

Schütz, Gabriela, 5, 6

Science and scientific literacy, 232–33, 243

Scotland, 17, 32, 179, 234, 248, 250

SEC. *See* Surveys of Enacted Curriculum

Secondary school. *See* Grade level